HAWAIIAN *Tropical* FLOWER ARRANGING

A Step-by-Step Guide

To Verena,
Happy arranging!
aloha,
Laurie Ide

HAWAIIAN *Tropical* FLOWER ARRANGING

A Step-by-Step Guide

Laurie Shimizu Ide

Library of Congress Catalog Card Number: 2004102401

First Printing, March 2004
1 2 3 4 5 6 7 8 9

ISBN 1-56647-663-1

All photographs by Barney Bareng and Laurie Shimizu Ide
Design by Mardee Domingo Melton

MUTUAL PUBLISHING
1215 Center Street, Suite 210
Honolulu, Hawai'i 96816
Ph: (808) 732-1709 Fax: (808) 734-4094
e-mail: mutual@mutualpublishing.com
www.mutualpublishing.com

Printed in Korea

Table of Contents

Acknowledgments . . . . **vi**

Preface . . . . **vi**

Introduction **1**

Flower Care & Handling **2**

Specific Flower Care Information . . **2**

Flower & Foliage Identification . . . **3**

Floral Equipment **9**

Containers **10**

Principles & Elements of Design . . **11**

Color Wheel **12**

Foliage Techniques **13**

American Style Arrangements **14**

Triangle (Symmetrical) Design **15**

Radiating Triangle Design **17**

Right & Left Triangle Design **19**

Scalene Triangle Design **21**

Vertical Design **23**

Crescent ("C" Curve) Design **24**

Hogarth ("S" Curve) Design **25**

Round Design **26**

Low and Long (Oval) Design **27**

European Style Arrangements **29**

Grouping **30**

Parallelism **31**

Framing **32**

Layering **33**

Landscape **34**

New Convention **35**

Colonial **36**

Pavè **37**

Biedermier **38**

Flemish **39**

Hand-Tied **40**

Oriental Style Arrangements **42**

Basic Guidelines in Japanese Flower Arranging **43**

Shoka **43**

Modern Rikka (Shimputai) **45**

Heika **48**

Moribana **49**

Basic Upright-Style Nageire **51**

Basic Slanting-Style Nageire **54**

Juuka or Jiyuka **56**

Bibliography **58**

ACKNOWLEDGMENTS

Mahalo

Mom and Dad

Elaine Mezurashi

Barney Bareng

Mr. & Mrs. Yuki Toguchi

Olomana Tropicals —
Ken and Lisa Vinzant

Sunset Tropicals —
Terrie & Patty

Maui Kula Lavender —
Lani & Ali`i

Waihale Products —
Alvin Tsuruda

Heidi Bornhorst

Dr. Richard Criley

Dr. Heidi Kuhnle

Judy Dela Cruz

Floral Resources/Hawaii, Inc.

Paradise Florist

Flo's Min Florist

Mylene's Floral Wholesale

Watanabe Floral

Aiea Florist

Photo Trends – Pearlridge

Mutual Publishing

And a special "thank you"
to Karl, Karley & Kamren

Preface

This book is designed for the flower arranging beginner. Step-by-step instructions are given to create 28 of the most popular classical and contemporary designs. Basic principles and techniques are covered.

This book is based on an informal way of teaching. Many years of learning, training, and practicing have been eliminated; a simple, short, and fun approach is the main goal. It allows for new ideas and changes in styles, stimulating creativity and design capabilities. Many adaptations and modifications are possible. You can create many new and different designs by adding to and mixing designs. Current trends and styles continue to add to the many techniques of floral design.

Quantities of flowers and foliage may vary, depending on the container and flower size. The flower placement order may change as long as the final result is the same. The flowers shown may be substituted with other flowers, following the same basic structure. To practice your arranging skills, artificial flowers may be substituted, creating an everlasting display.

This book is based on past experiences as a floral designer, on the job designing training, attending floral design school, taking workshops, classes, and reading on the subject matter. Information given may have oversights or be outdated. The methods and principles described in this book are to the best of the author's knowledge.

Introduction

With our tropical rains and warm sunshine, Hawai'i is home to the most beautiful flowers of the world. Their exotic, vibrant, and long-lasting quality adds excitement to an ordinary arrangement. It becomes dramatic and elegantly contemporary.

Today, more and more people are discovering the joy of flower arranging. You will find that once you start arranging flowers, your spirits will soar. Your home will come alive with a wonderfully refreshing atmosphere.

This book demonstrates and guides you as you create your own dramatic design, using tropical flowers. It guides you through the creation of 28 popular floral designs and techniques. Flower arranging is fun and simple when following the easy step-by-step, do-it-yourself instructions, accompanied by self-explanatory photos. Instructions are given to create American style flower arrangements, European style arrangements and designing techniques, and the harmonious Oriental style arrangements. Anyone can create beautiful flower arrangements. This book will demonstrate how easy it is. The more you practice, the better you'll get.

As you create your own arrangements, you will learn flower and foliage identification, the elements and principles of floral design, design mechanics, care and handling techniques, and floral designing skills. Many valuable tips like wiring, cutting, and bending the stems are covered. An in-depth coverage of tools and supplies used by professional florists and how to select and care for the freshest flowers for the longest vase life are also included.

Flower Care & Handling

PICKING: Pick flowers from your garden in the early morning, before the midday sun has struck. Use a tall bucket filled with cool water. Place flowers in the water immediately, allowing them to absorb water before the stems begin to heal.

PURCHASING: Purchase the freshest flowers available. Avoid buying flowers with discolored foliage or stems. Ask for the proper care tag and a packet of floral preservative.

RE-CUTTING: Re-cut about 1" off the stems when you get home. Cut it at a 45-degree angle, submerged in cool water. A 45-degree cut allows a greater drawing area for water to be absorbed. Cutting under water will ensure that no air bubbles get trapped in the stem, blocking the water flow up to the flower. Very thick stems, too thick to cut with shears, may be cut with a small saw, or pruner, or crushed with a hammer, about 1" from the bottom.

FOLIAGE REMOVAL: Remove all foliage that will fall below the water line. It reduces the rotting of leaves producing bacteria.

WATER: Place flowers in a tall container filled with cool water and add floral preservatives (optional). If no floral preservative is used, a drop of bleach in the water will help to prevent the cloudiness caused by bacterial growth. Add water to the container daily as needed.

SHINE: Apply an aerosol leaf shine to heliconias, bird-of-paradise bracts, and foliage. This adds to their vibrancy and increases their life span.

LOCATION: Place flowers in a cool location. Keep away from direct sunlight, drafts, and extreme hot or cold temperatures, causing premature wilting and shortening the life of the flowers. Flowers should not be placed near fruits and vegetables. Many fruits and vegetables give off ethylene gas, which shortens the flowers' life.

MAINTENANCE: Check the water level daily; add if needed. If you do not use a floral preservative, change and fill the water every two days. Rinse the stems from any bacteria growth and re-cut under water about 1" from the bottom. Clean the container and refill the water. Remove dying flowers and foliage in order to preserve the fresher flowers. This will not only improve the appearance of the arrangement, but also remove the decaying materials that cause bacteria build-up. If a special cut-flower preservative is used, just add water to the container daily. Replenish the solution every three to four days. If arranging foam is used, make sure it does not dry out. By following these basic proper flower care and handling guidelines, you can enjoy the maximum vase life of the flowers and the natural beauty they bring.

Specific Flower Care Information

ANTHURIUM: Anthuriums are prized for their especially long vase life and vibrant colors. Hawai'i-grown anthuriums are available in colors such as red, orange, pink, white, green, and purple, not even mentioning the Obakes (ghost flowers, with a touch of green), the speckled-type and the tulip anthuriums. Purchase anthuriums with strong, hard necks; a weak neck indicates an immature flower. Anthuriums are mature when their spadix changes color from a yellowish/off-white to white with a rough rather than smooth surface. An immature flower will not last as long as a mature flower. Place in a tall, clean container of cool water, with or without floral preservatives. Sprinkle the flower heads with cool water daily to prevent dehydration. Let stand for at least 30 minutes before arranging. Do not refrigerate. Repeat this process every 3 to 5 days. For wilted flowers, submerge the whole flower, including the stem, in water for 30 minutes. They will drink more through their heads than through their stems. Depending on the variety, anthuriums may last up to 28 days.

BIRD-OF-PARADISE: Birds-of-paradise are prized for their exotic, bright orange, bird-like beauty. Birds-of-paradise are available year-round, less during the winter months. If your bird-of-paradise is in a bud stage, submerge the head in cool water for 15 seconds. Reach into the bract with your thumb, pulling out one or two sets of petals. Rub an aerosol leaf shine on to the bracts to remove the seeping nectar, adding luster and vibrancy to the blossom. Do not apply the leaf shine to the petals, to prevent discoloration. Do not refrigerate. Birds-of-paradise may last up to 14 days.

GINGER: Gingers are prized for their bright colors and exotic shapes. Many gingers are available year-round. Trimming off the leaves will prolong the life of the flower. Do not refrigerate. Gingers may last up to 14 days.

HELICONIA: Heliconias are prized for their bright colors, exotic shapes, and bold lines. Many heliconias are available year-round. Trimming off the leaves will prolong the life of the flower. Remove the flowers inside the bracts to minimize odor and decay. Wash out the bracts in water with a little detergent and rinse. Soak the bracts in water for 30 minutes every 3 days for a longer vase life. Do not refrigerate. Heliconias may last up to 14 days.

ORCHID: Orchids are prized for their exotic tropical island association. Orchids blossom from the bottom of the stem to the top. Purchase stems with the flowers fully open. Most orchids will not open once they have been cut. Spray-mist the flowers and pluck off spent flowers daily. Repeat and change the water every 2 to 3 days. If flowers appear wilted, submerge the whole flower, including the stem, in cool water for 30 minutes. They will drink more through their heads than through their stems, refreshing them as well.

DENDROBIUM: Dendrobium orchids are prized for their long arching, cascading sprays. Dendrobium orchids may last up to 14 days. Do not refrigerate.

ONCIDIUM: Oncidium orchids are prized for their bright yellow color and long arching, cascading sprays. Oncidium orchids may last up to 10 days. Do not refrigerate.

CYMBIDIUM: Cymbidium orchids are prized for their long lasting qualities. Refrigerate to prolong the aging process. Cymbidium orchids may last up to 14 days.

PHALAENOPSIS: Phalaenopsis orchids are prized for their cascading elegant sprays. Refrigerate to prolong the aging process. Phalaenopsis orchids may last up to 10 days.

CATTLEYA: Cattleya orchids are prized for their individual beauty. Refrigerate them to prolong the aging process. Cattleya orchids may last up to 5 days.

EPIDENDRUM: Epidendrum orchids are prized for their blooming qualities and dainty size. Refrigerate to prolong the aging process. Epidendrum orchids may last up to 14 days.

PROTEA: Proteas are prized for their extended vase life, exotic forms, textures, and subsequent use as a dried flower. Each variety has its own season. Proteas are thirsty flowers. Check water supply daily and change every 2 to 3 days. A floral preservative is recommended. Refrigerate to prolong the aging process. Some discolorations of the leaves are normal for such varieties as the minks and the duchess. Commercial floral preservatives, ideal storage temperatures, and adequate light help to lessen the discoloration of the leaves. Clip off the discolored leaves before arranging. Proteas may last up to 30 days, depending on the variety. If flowers appear wilted, submerge the whole flower, including the stem, in cool water for 30 minutes. They will drink more through their heads and leaves than through their stems.

FOLIAGE: Foliages require the same care as tropical flowers. Place in room-temperature water for 30 minutes before arranging. Place 1 to 2 drops of household bleach into the water. This will kill the bacterial growth in the water.

Flower & Foliage Identification

The following chart identifies many of the popular tropical flowers and foliage used in arrangements found in Hawai‘i. Common plant names are in quotes, while cultivar and variety names are in single quotes.

'Ozaki'
Light Red Anthurium
Anthurium andraeanum

'Nitta Orange'
Orange Anthurium
Anthurium andraeanum

'Marion Seefurth'
Pink Anthurium
Anthurium andraeanum

'Midori'
Green Anthurium
Anthurium andraeanum

'Tropical Mist'
White Anthurium
Anthurium andraeanum

'Exotica Speckled Red'
Speckled Anthurium
Anthurium andraeanum

'Mickey Mouse'
Obake Anthurium
Anthurium andraeanum

'Anuenue'
Rainbow Obake Anthurium
Anthurium andraeanum

'Kalapana'
Obake Anthurium
Anthurium andraeanum

'Oshiro Obake'
Obake Anthurium
Anthurium andraeanum

'Lady Jane'
Tulip Anthurium
Anthurium x antioquiense

'ARCS'
Tulip Anthurium
Anthurium hybrid

'Lavender Lady'
Tulip Anthurium
Anthurium hybrid

'Red Hot'
Tulip Anthurium
Anthurium hybrid

'Calypso'
Tulip Anthurium
Anthurium hybrid

'Lady Ruth'
Tulip Anthurium
Anthurium hybrid

'Sassy'
Parakeet Heliconia
Heliconia psittacorum

"Red Parakeet"
Parakeet Heliconia
Heliconia psittacorum

'Yellow Parrot'
Parrot Heliconia
Heliconia psittacorum x spathocircinata

'Tropica'
Parrot Heliconia
Heliconia psittacorum x spathocircinata

'Yellow Fan'
Heliconia
Heliconia lingulata

'Red Christmas'
Christmas Heliconia
Heliconia angusta

'Richmond Red'
Heliconia
Heliconia caribae x bihai

'Jacquinii Claw'
Heliconia
Heliconi bihai

"Lobster Claw"
Heliconia
Heliconia humilis 'Jacquin'

"Yellow Caribaea"
'Cream' Heliconia
Heliconia caribaea

"Red Caribaea"
Heliconia
Heliconia caribaea

'Rostrata'
Hanging Heliconia
Heliconia rostrata

'Sexy Pink'
Hanging Heliconia
Heliconia chartaca

"Red Collinsiana"
Hanging Heliconia
Heliconia collinsiana

"Bird-of-Paradise"
Strelitzia reginae

"White Bird-of-Paradise"
Strelitzia nicolai

"Red Ginger"
Ginger
Alpinia purpurata

'Eileen McDonald'
Pink Ginger
Alpinia purpurata

"Red Torch"
Ginger
Etlingera elatior

"Indonesian Wax"
Ginger
Tapeinochilus ananassae

"Shell Ginger"
Ginger
Alpinia zerumbet

"Beehive Ginger"
Ginger
Zingiber spectabile

"Shampoo Ginger"
Ginger
Zingiber zerumbet

"Turmeric/Olena"
Ginger
Curcuma domestica

"Green Rose"
Ginger
Costus maculatus

"Indian Head"
Ginger
Costus spicatus

'Barbatus'
Ginger
Costus barbatus

Blue "ginger"
"Ginger"
Dichorisandra thysiflora

"Mauve Dancing Lady"
Ginger
Globba winitti

"Rattlesnake"
Calathea
Calathea crotalifera

"Cigar"
Calathea
Calathea lutea

"Green Ice"
Calathea
Calathea cylindrica

"Lavender Supreme"
Dendrobium Orchid
D. Jaq. Thomas 'Uniwai Supreme'

"Pearl White"
Dendrobium Orchid
D. Jaq. Hawai'i 'Uniwai Pearl'

"Royal Purple"
Dendrobium Orchid
D. Jaq. Thomas 'U. prince/princess'

"Blush White"
Dendrobium Orchid
D. Jaq. Thomas 'Uniwai Blush'

"Red Soniya"
Dendrobium Orchid
Dendrobium hybrids

'Oncidium'
Orchid
Oncidium 'Grower Ramsey'

'Sherry Baby'
Oncidium Orchid
Oncidium hybrid

'Epidendrum'
Orchid
Epidendrum species

"White Phalaenopsis"
Orchid
P. (Yukimai x Pinlong Spring) x Dtps. Casablanca Jey

"Pink Phalaenopsis"
Orchid
P. Dtps. Formosa Rose 'B#2' x P. Brother/ Purple 'Brother'

"Light Green Cymbidium"
Orchid
Cymbidium hybrid

"White Cymbidium"
Orchid
Cymbidium hybrid

"White Cattleya"
Orchid / *C. LC Shellie Compton 'Touch of Class' x BLC Meditation 'Queen's Dowry'*

"Lavender Cattleya"
Orchid
C. BLC. Mahina Yahiro 'Ulii' AM AOS

"King"
Protea
Protea cynaroides

'Mayday'
Protea
Protea hybrid magnifica x neriifolia

'Susara'
Protea
Protea magnifica x sussanae

"Duchess"
Protea
Protea eximia

'Jester'
Protea
Protea obtusifolia

'Pink Ice'
Protea
Protea hybrid

"White Mink"
Protea
Protea neriifolia

"Pink Mink"
Protea
Protea neriifolia

"White Owl Mink"
Protea
Protea neriifolia

'Rose Mink'
Protea
Protea laurifolia

'Hapapa Pink Mink'
Protea
Protea neriifolia

'Naiobe'
Protea
Protea neriifolia x laurifolia

"Princess"
Protea
Protea grandiceps

"Honey Glow/Sugar Bush"
Protea
Protea repens

'Scarlet Ribbon'
Protea
Banskia coccinea

'Raspberry Frost'
Banskia
Banskia menziesii

'Orange Frost'
Banskia
Banskia prionotes

"Mint Julep / Rickrack"
Banskia
Banskia speciosa

'Integrifolia'
Banskia
Banskia integrifolia

"Golden Dryandra"
Banskia
Dryandra fomosa

'Indian Summer'
Banskia
Banskia occidentalis

"Hairpin Banskia"
Banskia
Banskia spinulosa

"Silver Tree"
Leucadendron
Leucadendron argenteum

'Hawai'i Sunbeam'
Pincushion Protea
Leucospermum hybrid

"Yellow Sunburst"
Pincushion Protea
Leucospermum cordifolium

'Scarlet Ribbons'
Pincushion Protea
Leucospermum cordifolium

"Red Rocket"
Pincushion Protea
Leucospermum reflexum hybrid

"Yellow Rocket"
Pincushion Protea
Leucospermum reflexum
'Brandy Dela Cruz'

'Safari Sunset'
Leucadendron
Leucadendron

"Safari Green"
Leucadendron species

"Plumosum"
Leucadendron
Leucadendron rubrum

"Yellow Flame Tip"
Leucadendron
Leucadendron salignum

"Lavender Banana"
Musa ornata
Musa ornata

"Miniature Pineapple"
Ananas comosus

"Ti Leaf "
Cordyline fruticosa

'Monstera'
Monstera deliciosa

"Variegated Pandanus"
Pandanus veitchii

Sago "palm"
Cycas revoluta

"Areca Palm"
Chrysalidocarpus lutescens

'Raphis Palm'
Rhapis excelsa

"Dwarf Brassaia"
Shefflera arboricola

"Chinese Evergreen"
Aglaonema hybrids

"Varigated Croton"
Codiaeum variegatum

"Horsetail"
Equisetum hyemale

Goza "grass"
Junkus effusus

"Giant Papyrus"
Cyperus papyrus

"Dwarf Papyrus"
Cyperus isocladus

"Umbrella Papyrus"
Cyperus alternifolius

Ming "fern"
Asparagus macowanii

"English Ivy"
Hedera helix

"Pothos Vine"
Epipremnum aureum

'Lycopodium'
Lycopodium cernuum

"Leather Fern"
Arachnoides andiantiformis

"Rabbit's Foot Fern"
Davallia fejeensis

Asparagus "fern"
Asparagus plumosus

"Laua'e Fern"
Microsorium scolopendria

"Sword Fern"
Polystichum munitum

Sprengeri "fern"
Asparagus densiflorus

Floral Equipment

Proper tools and supplies are important as with all trades and hobbies. A few basic tools and supplies are necessary before you begin flower arranging. Floral equipment is categorized into four general categories: 1) tools, 2) supplies, 3) mechanics, and 4) containers.

TOOLS

KNIFE: A sharp knife is the most important tool in flower arranging. It enables a clean horizontal cut of the stem allowing the maximum amount of water flowing up to the flower.

FLORAL SHEARS: Floral shears are used for stems and branches that are too thick to trim with a knife. It is used to trim stems under water.

PRUNING SHEARS: Pruning shears are used to cut thick, hard, woody stems that are difficult to cut with a knife or floral shears.

WIRE CUTTER: Wire cutters are used to cut wire and chicken wire.

SCISSORS: Scissors are used to trim leaves and cut other materials.

NEEDLE PIN HOLDER STRAIGHTENER: The needle pin holder straightener is used to repair bent needles and remove debris.

GLUE GUN: A glue gun dispenses hot glue that sets just as fast as it takes the glue to cool. Extreme care should be taken when using a glue gun to prevent burns and blisters.

WATERING CAN: A long-spout watering can is used to add water to the arrangement without wetting the container.

WATER SPRAYER (MISTER): A water mister enables a fine mist of water on the flowers and foliage for added vase life.

WATER BOWL: A water bowl is used to give the stems an underwater cut while arranging. An added strainer helps to prevent the stems from going into the drain.

SUPPLIES

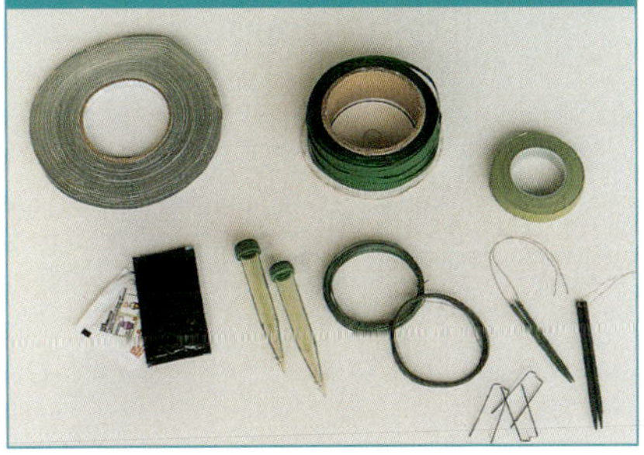

FLORAL ADHESIVE TAPE: Floral adhesive tape is a waterproof tape used in flower arranging to secure the floral foam to the container.

FLORAL TAPE: Floral tape is stretchable tape that secures and camouflages stems, floral wire, and picks in place.

TWIST WIRE: Twist wire is used to secure the liner and floral foam to a basket.

WIRE (#22 & #24): Floral wire is used to bind, mend, and bend stems.

WATER PICKS: Water picks are plastic vials with rubber caps. They provide a water supply when the stems are not long enough to reach the water level.

WIRED PICKS (3" & 4"): Wired picks are wooden picks with wire attached to one end. The wire is used to fasten the flower stem to the pick.

GREENING PINS: Greening pins help to secure moss or foliage to the floral foam.

FLORAL PRESERVATIVES: Commercial floral preservatives prolong the vase life of flowers. They are available from a florist or floral supply store.

MECHANICS

FLORAL FOAM: Floral foam comes in dried blocks that can be cut to fit the container. Complete saturation of the foam is important. There are various types of foam. Because of their weight, tropical flowers require the strongest floral foam available. Soak the foam in water for about 10 minutes until thoroughly saturated. Floral foam should be positioned so that about 1 to 3" extend above the container, depending on the flowers, the size of the container, and the size of the arrangement. Be sure there is room to add water and hold an adequate water supply. When arranging, insert flower stems securely into the foam. Stems should be inserted at least $1^1/_2$ to 3" into the foam depending on the foam size. If inserted incorrectly, re-cut your stem and place it in a different spot to allow for better stability and better water absorption. Add water to the container daily. Do not use dried foam after it has been soaked—it will not absorb water correctly.

CHICKEN WIRE: Chicken wire is used to provide extra support over the floral foam when using heavy flowers or large quantities of flowers. It is also used in tall opaque containers as a holding medium. Select the 1" mesh netting, which is pliable and easy to use. To avoid injury from the sharp wire edges, cut the mesh netting smaller than the size of the floral foam.

NEEDLE PIN HOLDER (FROG): A needle pin holder consists of sharp pins, closely spaced and anchored to a heavy base. It is also called a Kenzan, meaning "mountain of sword points." It is used in Oriental designs and is available in many sizes and shapes.

PEBBLES: Pebbles are used to secure, raise, and cover the needle pin holder in the container of an Oriental style arrangement. They are also used in natural landscaping designs

Containers

A flower container is a water-holding vessel in which flower stems can be arranged. Select the right container that will harmonize with each floral arrangement.

Containers may be made of ceramic, clay, wood, metal, porcelain, stoneware, glass, crystal, wicker, bamboo, gourd, or plastic. They come in different sizes, shapes, colors and finishes.

Before selecting a container, know the colors and textures of the background in which the arrangement will be placed, degree of formality of the occasion, and the purpose for the arrangement. Then decide on the size, shape, color, texture, and formality of the container. The container should be large enough to hold the materials without crowding and deep enough to hold sufficient water.

As you master the balance and composition of floral designing, you will find that almost any container may be used for almost any flower arrangement. However, as a beginner, choose containers of good design that will help ease the learning process instead of making it more challenging.

(a) *Compote* **(b)** *Pitcher* **(c)** *Bud Vase* **(d)** *Pot* **(e)** *Rectangular Pillow* **(f)** *Novelties* **(g)** *Basket* **(h)** *Cylinder* **(i)** *Mug* **(j)** *Urn* **(k)** *Jardinière* **(l)** *Chalice* **(m)** *Bowl* **(n)** *Tray*

Principles & Elements of Design

All flowers are beautiful when cut and casually placed in a container. However, these same flowers arranged into a striking design can be a work of art. This can be achieved by following these basic principles and elements of design.

Principles of Design

BALANCE: Balance is most important in all flower arrangements. It may be symmetrical or asymmetrical. It gives the feeling of security and stability. It can be mechanical or visual. Mechanically, the arrangement should not look as if it is about to fall over. Visually, balance is achieved by proper use of color, texture, size, and space. The stems should appear to originate from a central or "radial" point in the arrangement. A radial point or "binding point" is the imaginary spot inside a container where all the stems meet, creating balance. Many European and contemporary arrangement styles have more than one radial point. Balance between the arrangement and the container is usually controlled by the longest line in the arrangement. This one line will balance the mass of the container. For this reason, a simple arrangement using one or very few stems will balance with the container. If this one line is not well proportioned, the balance will be off, no matter how many more stems are added.

FOCAL POINT OR DOMINANCE: The focal point is important to anchor the design. It dominates the composition. This area is the center of interest in an arrangement. It is the point to which the eye is drawn.

SCALE AND PROPORTION: Scale and proportion is the relationship of the materials to each other in size, shape, quantity, setting, as well as to the container. There should be harmony between the flowers and the container, and between the arrangement and its surroundings.

RHYTHM: Rhythm is the visual path of repetition of line, color, mass, or mood of the materials in the arrangement. It creates the feeling of direction and a flowing, graceful movement. Crossing or misplacing stems breaks the visual path, disrupting the rhythm of the design. However, the crossings of the stems are acceptable in freestyle designs.

HARMONY: Harmony exists when all parts of a floral design are flowing together. The container, the flowers, the foliage, and composition should have something in common with each other.

Elements of Design

COMPOSITION: Composition is the main guiding element of a floral design. Before starting an arrangement, consider whether the final setting of the display calls for a formal, modern, or casual theme. Then think about the container, flowers, foliage, size, shape, line, colors, mass, accessories, voids, texture, and background of the arrangement. Good composition relates all of these parts to achieve harmony.

LINE: Line is created by the use of materials placed in a predetermined linear pattern. It can be curving and graceful, as well as straight and dynamic. It provides the shape and structure of a floral arrangement. It is the visual path that the eye follows. Direct lines are actual lines that are easy for the eyes to follow. Indirect lines are created by a series of repetitious materials throughout the design.

FORM: Form is the three-dimensional shape of a floral design. It may be circular, triangular or curved.

SPACE: Space in a design can be positive or negative. Positive space is the area within a floral design that is occupied by the plant material or accessories. Negative space is represented by planned emptiness or "void" in the arrangement. Both negative and positive spaces are important to the design.

TEXTURE: Texture refers to the surface qualities of plant and flower materials, accessories, and container. Qualities such as smoothness, roughness, glossiness, delicateness, fuzziness, brightness (or dullness), and size of leaves add character and diversity to a flower arrangement. The mixture of flowers, with different textures, increases depth and visual interest.

DEPTH: Depth is the variations of the length of stems arranged from back to front or center outward. It prevents the arrangement from looking flat.

MASS: Mass refers to a technique that groups materials together. It usually creates a focal point in the arrangement. This technique may be used to tuck materials deep into the arrangement to add balance, texture, and depth.

COLOR: Color is an important element in floral designing. It is the harmonious blending of color that will affect the mood and feeling in which an arrangement is viewed. The 12 part color wheel shows an orderly arrangement of the basic colors: primary, secondary, and tertiary, curved into a circle.

Color Wheel

There are six main color harmonies in floral design: monochromatic, analogous, complementary, split complementary, triadic, and tetrad or double split complementary.

PRIMARY COLORS: Primary colors are red, yellow, and blue. Primary colors cannot be mixed from any other colors

SECONDARY COLORS: Secondary colors are orange, green, and violet. They are mixtures in equal amounts of any two primary colors.

TERTIARY COLORS: Tertiary colors are red-orange, yellow-orange, yellow-green, blue-green, blue-violet, and red-violet. They are produced when mixtures of primary and adjacent secondary colors are combined in equal amounts.

Color Harmonies

MONOCHROMATIC: Monochromatic is the use of all flowers in an arrangement having tints, shades, and tones of one color.

ANALOGOUS: Analogous colors are those colors found next to each other on the color wheel.

COMPLEMENTARY: Complementary colors are those opposite each other on the color wheel.

SPLIT COMPLEMENTARY: Split complementary colors are those two colors directly adjacent to its complementary color.

TRIAD: Triad color harmony combines three colors at equal distances from each other on the color wheel.

TETRAD OR DOUBLE-SPLIT COMPLEMENTARY: Tetrad color harmony combines colors four hues equally spaced on the color wheel.

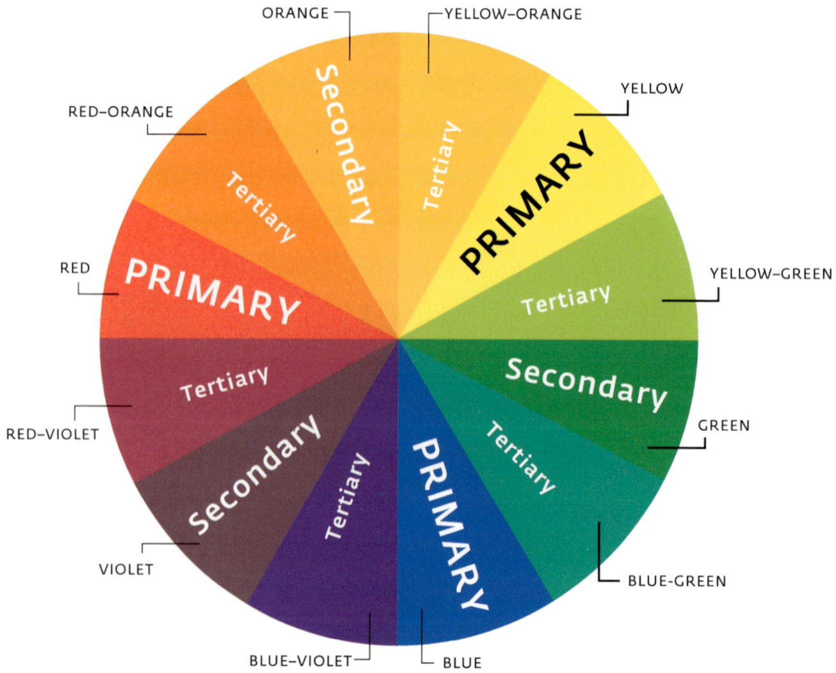

Foliage Techniques

Leaves may be altered by creative techniques that change their form. Shown below are some examples.

TRIM:

1

2

3

4

1. Trim a ti leaf.

2. Trim an areca palm frond.

3. Trim a sago palm leaf.

4. Trim off the edges of a raphis palm in a ragged pattern.

CUT:

5

6

7

5. Cut a ti leaf to create a unique shape.

6. Cut an areca palm frond to create a unique shape.

7. Cut a sago palm frond to create a unique shape.

SLICE:

8

8. Slice or tear a ti leaf diagonally following the vein pattern. Gather and secure the base to the main stem with wire or floral tape.

LOOP:

9

10

11

12

9. Loop and pierce a ti leaf stem into the middle of the leaf to create a "bow."

10. Loop or roll over a ti leaf tip to create a loop. Secure it with staples or tape.

11. Loop, trim, and secure a ti leaf onto a wired pick to create a "bow."

12. Loop, trim, and secure a money tree leaf onto a wired pick to create a "bow."

LAYER:

13

13. Layer three ti leaves, trim, and secure them on a wired pick.

KNOT:

14

14. Knot the end of a pandanus leaf.

AMERICAN *Style*

ARRANGEMENTS

American style designs are composed of primary flowers creating lines to form shapes. Secondary and tertiary flowers are then added to the primary flowers. There are 10 main designs: triangle, radiating triangle, right and left triangle, scalene triangle, vertical, crescent, hogarth or "S" shape, round, and oval, which are the main design line forms.

Triangle (Symmetrical) Design

MATERIALS:

- **1 Container – with secured, wet foam**
- **12 Red gingers**
- **10 Anthuriums**
- **24 Ti leaves**
- **10 Lycopodium stems**
- **12 Leather ferns**

CONTAINER PREPARATION:

Cut a piece of well-soaked floral foam to fit inside the container, leaving about 1/2" for water on both sides of the foam. The foam should extend about 2" above the top of the container. Secure with waterproof floral tape. Add floral preservative and cool water. (Optional: add chicken wire before taping down foam for added support.)

Line flowers are positioned forming a triangle with three equal sides (equilateral), or two equal sides, with a different length for the base (isosceles). Filler flowers fill in the triangle shape. Use a medium-deep container.

1. Cut the straightest ginger 2 1/2 to 3 times the height (or width) of the container. Cut at a 45-degree angle. Insert it to the center back of the foam, about 2" from the edge. Cut 2 gingers for the sides to create a triangle shape. Allow an additional 3" for insertion allowance. Insert them on the sides, allowing the stems to rest on the rim of the container.

2. Insert 4 additional gingers, keeping within the triangle to create the total area space.

3. Insert the remaining 5 gingers within the triangle frame at even distances apart.

4. Select the straightest, nicest, tallest anthurium for the top. Select 2 anthuriums for the sides. Insert them between the gingers creating a smaller triangle.

5. Fill in the remaining 7 anthuriums between the gingers, within the triangle, at even distances apart.

6. Add about 9 ti leaves in the back of the arrangement, creating a triangle shape, to cover the mechanics. Trim the leaves if needed. See page 13 for foliage techniques.

7. Trim and add about 3 ti leaves to the front of the arrangement. Angle the leaves slightly downward.

8. Fill in the triangle with ti leaf bows and leaves to cover the foam. See page 13 for foliage techniques.

9. Add lycopodium stems between the flowers to add softness, create interest, and to cover the mechanics. Keep the lycopodium within the triangle shape.

10. Add ti leaf bows in the back of the arrangement to cover the mechanics.

11. Evaluate the design; adjust, if needed, to complete the "triangle" arrangement.

1

2

3

TRIANGLE (SYMMETRICAL) DESIGN *Continued from previous page.*

4

5

6

7

8

9

10

11

Radiating Triangle Design

MATERIALS:

- **1 Container – with secured, wet foam**
- **12 Obake anthuriums – medium to large**
- **24 Dendrobium orchids**
- **24 Ti leaves – medium size**
- **12 Rabbit's foot ferns**

CONTAINER PREPARATION:

Cut a piece of well-soaked floral foam to fit inside the container, leaving about 1/2" for water on both sides of the foam. The foam should extend about 1 1/2" above the top of the container. Secure with waterproof floral tape. Add floral preservative and cool water. (Optional: add chicken wire before taping down foam for added support.)

Line flowers are positioned forming a fan shape. The basic triangle steps are followed, however the outline flowers are extended outside the triangle shape to create a rounded fan-shaped form. Use a shallow to medium-deep container.

1. Cut the straightest, best anthurium stem about 3 times the width (or height) of the container. Cut at a 45-degree angle. Insert toward the center back of the foam, about 2" from the edge. Measure and cut 2 anthuriums longer than half the height of the arrangement. Insert them on both sides, allowing their stems to rest on the rim of the container, to create triangular shape.

2. Insert 4 additional anthuriums to establish the form, creating a fan shape.

3. Insert the remaining 5 anthuriums, within the fan-shaped frame, at even distances apart.

4. Add about 11 ti leaves in the back of the arrangement, creating a fan shape, to cover the mechanics. Trim the leaves if needed. See page 13 for foliage techniques.

5. Trim and add about 6 ti leaves to the front of the arrangement. Angle the leaves slightly downward.

6. Fill in the fan shape with ti leaf bows and leaves to cover the foam. See page 13 for foliage techniques.

7. Add ferns to cover anthurium stem mechanics. Keep the fern within the fan shape.

8. Add dendrobium orchids within the fan-shaped design.

9. Add ti leaf bows in the back of the arrangement to cover the mechanics.

10. Evaluate the design; adjust, if needed, to complete the "radiating triangle" arrangement.

1

2

3

4

5

6

RADIATING TRIANGLE DESIGN *Continued from previous page.*

7

8

9

10

Right & Left Triangle Design

MATERIALS:

- **1 Container – medium-deep with secured, wet foam**
- **12 Anthuriums – large Obake**
- **30 Ti leaves – medium**
- **24 Lycopodium stems**

CONTAINER PREPARATION:

Cut a piece of well-soaked floral foam to fit inside the container, leaving about 1/2" for water on both sides of the foam. The foam should extend about 1 1/2" above the top of the container. Secure with waterproof floral tape. Add floral preservative and cool water. (Optional: add chicken wire before taping down foam for added support.)

Line flowers are positioned forming the shape of a half of a symetrical triangle. It is an asymmetrical triangle. Flowers for a right-angled arrangement are positioned in a 90-degree right angle from the top flower to the right bottom flower. Likewise, the flowers for a left-angled arrangement are positioned in a 90-degree left angle from the top flower to the left bottom flower. These two arrangements form a symmetrical triangle when put together.

RIGHT-TRIANGLE ARRANGEMENT

1. Cut the straightest, best anthurium about 3 times the width (or height) of the container. Cut at a 45-degree angle. Insert it to the left back of the foam, about 2" from the edge. Cut and insert 2 anthuriums for the bottom sides of the arrangement, keeping the right flower stem longer than the left side, creating a right triangle.

2. Cut and insert 2 anthuriums for the topsides. Position the right flower slightly higher than the left flower. Keep within the right triangle shape.

3. Add 4 anthuriums to the sides, at even distances apart, to create a right triangle shape.

4. Fill in the remaining 3 anthuriums within the triangle shape, at even distances apart.

5. Add trimmed ti leaves to create a frame in the back and the front bottom of the arrangement, forming a right triangle shape to cover the mechanics. Fill in the triangle with ti leaf bows to cover the foam. See page 13 for foliage techniques.

6. Add lycopodium stems between the flowers to add softness, create texture, and to cover the mechanics. Keep the lycopodium within the triangle shape.

7. Add ti leaf bows in the back of the arrangement to cover the mechanics.

8. Evaluate the design; adjust, if needed, to complete the "right triangle" design. Shown: "left and right triangle" arrangements.

1

2

3

RIGHT & LEFT TRIANGLE DESIGN *Continued from previous page.*

4

5

6

7

8

Scalene Triangle Design

MATERIALS:

- 1 **Container – medium-deep with secured, wet foam**
- 10 **Christmas heliconias**
- 7 **Anthuriums – medium**
- 11 **Dendrobium orchids**
- 12 **Ti leaves – medium**
- 12 **Leather ferns**
- 5 **Raphis leaves**

CONTAINER PREPARATION:

Cut a piece of well-soaked floral foam to fit inside the container, leaving about 1/2" for water on both sides of the foam. The foam should extend about 1 1/2" above the top of the container. Secure with waterproof floral tape. Add floral preservative and cool water. (Optional: add chicken wire before taping down foam for added support.)

Line flowers are positioned forming the shape of a scalene or asymmetrical triangle. All three sides of the triangle are unequal. The line of this design can extend either to the right or the left. This triangular shape creates a styled appearance. Demonstrated is a tall, right scalene triangle design.

1. Select the straightest, best heliconia. Cut the length 3 times the width (or height) of the container. Cut at a 45-degree angle. Insert it to the center back of the foam, about 2" from the edge. Cut and insert 2 anthuriums for the bottom sides of the arrangement, keeping the right flower stem longer than the left side, creating an asymmetrical triangle.

2. Insert 3 additional heliconias, keeping within the asymmetrical triangle shape.

3. Insert the remaining 4 heliconias within the triangle, at even distances apart.

4. Add in all 7 anthuriums between the heliconias, within the triangle, at even distances apart.

5. Add about 10 trimmed ti leaves to the back of the arrangement, creating a scalene-shaped triangle to cover the mechanics. Trim the leaves if needed. See page 13 for foliage techniques.

6. Trim and add about 2 ti leaves to the front of the arrangement. Angle the leaves slightly downward. Add about 5 trimmed, raphis palm leaves, evenly spaced within the triangle.

7. Add a touch of leather ferns to cover the foam mechanics.

8. Add the dendrobium orchids between the flowers to add softness, create interest, and to cover the mechanics of the arrangement. Keep the orchids within the triangle shape.

9. Add ti leaf bows in the back to cover the mechanics. See page 13 for foliage techniques.

10. Evaluate the design; adjust, if needed, to complete the "scalene triangle" arrangement.

1

2

3

4

SCALENE TRIANGLE DESIGN *Continued from previous page.*

5

6

7

8

9

10

Vertical Design

MATERIALS:

1 Container – long, narrow neck
3 Mickey Mouse Obake anthuriums – small
1 Curly willow branch
3 Areca palm leaves
1 Brassaia leaf – dwarf
2 Leather ferns
2 Chinese evergreen leaves

CONTAINER PREPARATION:

Add floral preservative and cool water.

A vertical design is also known as a line design. Line flowers are positioned creating a vertical line. Its form is narrow, extending straight upward. Vertical designs are popular in bud vases.

1. Select the tallest, straightest flower for the top. Place the remaining 2 flowers below the top flower in a line.

2. Trim all stems together to the length where the top flower is about 2 times the height of the vase. Insert all at once into the vase.

3. Insert 3 trimmed areca palm leaves in the back of the flowers for added linear foliage. Add 2 Chinese evergreen leaves to the left side of the flowers. Insert 2 leather ferns, one to the right and the other to the middle of the design.

4. Insert 1 brassaia leaf to the front rim of the vase. Add 1 stem of curly willow in the back of the areca leaves for added height and style.

5. Evaluate the design; adjust, if needed, to complete the "vertical" design.

1

2

3

4

5

Crescent ("C" Curve) Design

MATERIALS:

- 1 Container – with secured, wet foam
- 8 Dendrobium orchids
- 4 Cymbidium orchids
- 7 Parrot heliconias
- 8 Leather ferns
- 4 Brassaia leaves – dwarf
- 3 Water tubes – 4"
- 6 Wires – #24 gauge

CONTAINER PREPARATION:

Cut a piece of well-soaked floral foam to fit inside the container, leaving about 1/2" for water on both sides of the foam. The foam should extend about 1 1/2" above the top of the container. Secure with waterproof floral tape. Add floral preservative and cool water. (Optional: add chicken wire before taping down foam for added support.)

Flowers are used to create the shape resembling a crescent moon. This design is known for its graceful, curved lines creating motion. The curve of the crescent is usually narrow at the ends, becoming wider at the focal point. It is an asymmetrical arrangement with one of its sides higher than the other. The crescent arrangement may be arranged in a shallow or medium-deep, round container.

1. Wire the dendrobium orchid stems to create the curved shape (if needed).

2. Select and cut the strongest, best, curved dendrobium stem to the length 3 times the height (or width) of the container. Cut at a 45-degree angle. Insert it to the back left of the foam, about 1" from the edge. Cut and insert the remaining 5 dendrobium stems creating a crescent shape, 3 stems on each side.

3. Add 4 cymbidium orchids, in water tubes, evenly spaced apart. Place in a "C"-shape line to enhance the crescent design and create a focal point.

4. Add 7 heliconias within the crescent shape, evenly spaced apart.

5. Insert 4 brassaia leaves and leather fern to cover the front and back foam mechanics, keeping within the crescent-shape form.

6. Evaluate the design; adjust, if needed, to complete the "crescent" arrangement.

1

2

3

4

5

6

Hogarth ("S" Curve) Design

MATERIALS:

- 1 Container – with secured, wet foam
- 6 Dendrobium orchids
- 3 Cymbidium orchids
- 6 Anthuriums – Lady Jane
- 2 Laua`e leaves
- 6 Leather ferns
- 3 4" water tubes

CONTAINER PREPARATION:

Cut a piece of well-soaked floral foam to fit inside the container, leaving about 1/2" for water on both sides of the foam. The foam should extend about 1 1/2" above the top of the container. Secure with waterproof floral tape. Add floral preservative and cool water. (Optional: add chicken wire before taping down foam for added support.)

Flowers are used to create an "S" shape. It is a vertical "S"-curved line that extends below the rim of the vase. This line creates motion. Careful selection of a focal point flower, or a group of flowers, is needed for the center of the arrangement. Only a few flowers are required to create this hogarth design. A tall vase is recommended for this arrangement style.

1. Select the strongest, curved dendrobium stem. Cut it to 1 1/2 times the height of the container. Cut at a 45-degree angle. Insert it to the back left of the foam, about 1" from the edge. Cut and insert the remaining 5 dendrobium stems, creating an "S"-shape curve, 3 stems on each side. Wire the dendrobium orchid stems to create the curved shape if needed.

2. Add 3 cymbidium orchids, in water tubes, to create a focal point. Keep the flowers within the "S"-shaped frame.

3. Insert 6 anthuriums, evenly spaced apart, within the "S"-shaped form.

4. Add 2 laua'e leaves in the back of the arrangement to cover the mechanics following the "S" design.

5. Add leather fern to the front and back of the arrangement to cover the foam mechanics. Keep within the "S" form.

6. Evaluate the design; adjust, if needed, to complete the "hogarth" design.

1

2

3

4

5

6

Round Design

MATERIALS:

- **1 Container – medium-deep with secured, wet foam**
- **1 King protea**
- **2 Pink ice proteas**
- **2 Pink mink proteas**
- **4 Pincushion proteas**
- **8 Safari sunset leucadendrons**
- **12 Lycopodium stems**
- **6 Leather ferns**

CONTAINER PREPARATION:

Cut a piece of well-soaked floral foam to fit inside the container, leaving about 1/2" for water on both sides of the foam. The foam should extend about 1 1/2" above the top of the container. Secure with waterproof floral tape. Add floral preservative and cool water before taping down foam for added support.

Round arrangements are referred to as colonials, mounds, or centerpieces. It is a mass-type arrangement where flowers are put together creating a circular form. It can be viewed from all sides when it is placed in the center of a table. The general height rule for this design is 10" or less if it is to be placed in the center of a meeting table. The low height refrains from obscuring one's vision or inhibiting conversation. If the arrangement is for a coffee table or large room, there are no restrictions on its height.

1. Determine the height of the arrangement by the amount of flowers given. Cut the king protea to the desired height, plus 2 to 3" for foam insertion. Insert the king protea into the foam, facing straight up. Add 2 pink ice proteas on the sides of the king protea, creating an arch or mound.

2. Insert 2 pink mink proteas on the opposite sides of the king protea, creating an arch.

3. Add 4 pincushions in the spaces between the proteas, evenly spaced apart.

4. Add 4 safari sunsets below the pink ice and mink proteas with their stems almost resting on the rim of the container. Then, add the remaining 4 safari sunsets between the pink ice and mink proteas, evenly spaced apart.

5. Cut the lycopodium stems into shorter pieces. Fill in between the proteas, keeping within the circular form. This creates added color and texture in the design.

6. Add leather fern tips between the lycopodium stems to create softness and cover the foam mechanics. Keep within the circular form.

7. Evaluate the design; adjust, if needed, to complete the "round" design.

1

2

3

4

5

6

7

Low and Long (Oval) Design

MATERIALS:

1 **Container – rectangular or oval with secured, wet foam**
9 **Red gingers**
12 **Mickey Mouse Obake anthuriums**
10 **Dendrobium orchids**
24 **Ti leaves**

CONTAINER PREPARATION:

Cut a piece of well-soaked floral foam to fit inside the container, leaving about 1/2" for water on both sides of the foam. The foam should extend about 1 1/2" above the top of the container. Secure with waterproof floral tape. Add floral preservative and cool water. (Optional: add chicken wire before taping down foam for added support.)

The low and long design is also called a horizontal arrangement. Its form is low, long, and oval in shape. It is constructed like a line design, but flat and straight out on the table creating a centerpiece. It can be viewed from all sides when it is placed in the center of a table. The general height rule for this design is 8" or less if it is to be placed in the center of a meeting table or wedding head table. The low height refrains from obscuring one's vision or inhibiting conversation. If the arrangement is for a coffee table or a larger room, there are no restrictions on its height.

1. Determine the height, length, and width of your arrangement. Insert the top ginger in the middle of the foam to determine the height of the arrangement. Add 2 side gingers to determine the length. This creates your triangle form.

2. Add 4 gingers around the center ginger, evenly spaced apart, facing outward. Add 1 ginger to the front and 1 ginger to the back, resting on the lip of the container. This creates the width of the arrangement. Keep within the imaginary triangle shape.

3. Add 4 anthuriums around the top ginger, evenly spaced apart, keeping within the triangle frame.

4. Add 4 anthuriums, 2 on each side of the front and back ginger, keeping within the triangle frame.

1

2

3

4

LONG AND LOW (OVAL) DESIGN *Continued from previous page.*

5. Add 4 anthuriums, 2 on each side of the long, side gingers, keeping within the triangle frame. When completed, this frame looks oval in shape.

6. Add trimmed ti leaves under the arrangement, resting on the rim of the container. Create an oval shape. See page 13 for foliage techniques.

7. Add ti leaf bows to cover the foam; keep within the oval form.

8. Add 10 stems of dendrobiums between the gingers and anthuriums, at equal distances apart. Keep within the oval form.

9. Evaluate the design; adjust, if needed, to complete the "low and long" or "oval" design.

5

6

7

8

9

EUROPEAN *Style*

ARRANGEMENTS

European style designs often use an idea or "technique" creating a contemporary design. Once a technique is selected, a design is created using the technique in an interpretive style. Color, flower grouping, and natural simplicity are main points to remember when designing in the European style. Eleven popular design techniques and styles: grouping, parallelism, framing, layering, landscaping, new convention, colonial, pavè, Biedermier, Flemish, and hand-tied are shown.

Grouping

MATERIALS:

1 **Container – medium-deep with secured, wet foam**
7 **Parakeet heliconias**
6 **ARCS anthuriums**
2 **Pincushion proteas**
4 **Safari sunset leucadendrons**
1 **Purple banana**
2 **Dracaena leaves – small**
2 **Panax leaves**
3 **Croton leaves**
2 **Sago "palm" fronds**
5 **Leather ferns**

CONTAINER PREPARATION:

Cut a piece of well-soaked floral foam to fit inside the container, leaving about $\frac{1}{2}$" for water on both sides of the foam. The foam should extend about $1\frac{1}{2}$" above the top of the container. Secure with waterproof tape. Add floral preservative and cool water. (Optional: add chicken wire before taping down the foam for added support.)

Grouping is achieved by using the same variety of flowers in a specific section of the design in a group. Grouping showcases the individual flower varieties. It not only intensifies their colors, but also adds textural contrasts, providing a greater visual impact to the flowers. Note: This technique is not clustering, as clustering does not allow for space between the flower heads.

1. Insert 7 heliconias to the center back of the foam; keep them in a group at different heights. The height of the tallest flower should be about $2\frac{1}{2}$ times the height of the vase. Insert the banana flower to the front, right side of the heliconias.

2. Add 2 pincushions and 4 safari sunsets in a grouping formation.

3. Insert 6 ARCS anthuriums to the right and front side of the arrangement in a grouping pattern.

4. Add 2 panax leaves to the front and 2 dracaena leaves to the left side of the arrangement.

5. Continue to add foliage; 2 sago fronds to the right back and 3 croton leaves to the right front in groups.

6. Add leather fern to the front and back of the arrangement to cover the foam mechanics.

7. Evaluate the design; adjust, if needed, to complete the "grouping" design.

1

2

3

4

5

6

7

Parallelism

MATERIALS:

1 Container – rectangular with secured, wet foam
5 Red gingers
4 Parakeet heliconias
2 Rattlesnake calatheas
3 Costus gingers
4 Safari sunset leucadendrons
4 Mickey Mouse Obake anthuriums
4 Lycopodium stems
3 Leather ferns
1 Ming fern stem

CONTAINER PREPARATION:

Cut a piece of well-soaked floral foam to fit inside the container, leaving about 1/2" for water on both sides of the foam. The foam should extend about 1 1/2" above the top of the container. Secure with waterproof floral tape or hot glue. Add floral preservative and cool water.

The parallelism technique includes two or more groupings of flowers that are used in a repeated vertical pattern with space between each vertical group. There must be space between each vertical grouping of flowers to be defined as a pure parallel design. Each vertical grouping is usually created by one type of flower and should be of a different height than the grouping immediately next to it.

1. Insert 5 red gingers vertically at different lengths into the left side of the foam. Add 4 heliconias vertically at different lengths into the right side of the foam. Keep a void spacing between the 2 vertical lines.

2. Add 2 rattlesnake calatheas next to the base of the red gingers. Insert 3 costus gingers next to the base of the heliconias, keeping them in groups.

3. Insert 4 safari sunsets at the base of the red gingers. Add 4 anthuriums to the base of the costus gingers, keeping them in groups.

4. Add lycopodium stems to the back of the design to cover the mechanics, keeping in the vertical form. Cover the foam mechanics with leather and ming fern tips.

5. Evaluate the design; adjust, if needed, to complete the "parallel" design.

1

2

3

4

5

Framing

MATERIALS:

- **1 Container – round with secured, wet foam**
- **2 Torch gingers**
- **6 Red gingers**
- **3 Pandanus leaves**
- **3 Areca palm fronds**
- **2 Raphis palm fronds**
- **1 Laua'e leaf**
- **4 Leather ferns**

CONTAINER PREPARATION:

Cut a piece of well-soaked floral foam to fit inside the container, leaving about 1/2" for water on both sides of the foam. The foam should extend about 1 1/2" above the top of the container. Secure with waterproof floral tape. Add floral preservative and cool water. (Optional: add chicken wire before taping down the foam for added support.)

Framing is the technique when branches or foliage create a frame, showcasing a flower or group of flowers. Framing separates and calls attention to the flowers inside the frame.

1. Insert a tall, straight pandanus leaf into the center of the foam, facing to the right. Add the other pandanus leaf, about 3" shorter in length, facing to the left, into the right side of the foam. Cut a 1" vertical slit through the left pandanus, part way down the middle. Insert the tip of the right pandanus through the slit, creating a frame-like design.

2. Trim 3 areca palm fronds: 1 long, 1 medium, and 1 short in length. Insert the long frond first, backward, creating the left frame edge. Add in the medium frond to the left of the long frond, facing backward. Insert the short frond to the right, creating a horizontal plane.

3. Insert 2 gingers in a vertical pattern into the frame created. Use the nicest, smallest ginger on the top. Add 6 red ginger tips around the base of the lower ginger.

4. Add 2 laua'e leaves to the left of the design, creating the horizontal effect. Insert 2 trimmed raphis fronds to the front and back right of the gingers. Cover the foam mechanics, in the front and back, with leather fern.

5. Evaluate the design; adjust, if needed, to complete the "framing" design.

Layering

MATERIALS:

- **1 Container – round with secured, wet foam**
- **2 Birds-of-paradise**
- **2 Parrot heliconias**
- **3 Pandanus leaves**
- **13 Ti leaves**
- **4 Mickey Mouse Obake anthuriums**
- **2 Dendrobium orchid stems**
- **4 Ti leaves – small**
- **1 Ming fern stem**
- **1 3" wired pick**

CONTAINER PREPARATION:

Cut a piece of well-soaked floral foam to fit inside the container, leaving about 1/2" for water on both sides of the foam. The foam should extend about 1 1/2" above the top of the container. Secure with waterproof floral tape. Add floral preservative and cool water.

The layering technique is to form or arrange in layers. Materials are placed one on top of the other so they overlap.

1. Insert a tall, straight pandanus leaf into the center of the foam, facing to the right. Add 2 other pandanus leaves, about 3" shorter in length, facing to the left, into the right side of the foam. Cut 2 1" vertical slits through the left pandanus partway down the middle. Insert the tips of the right pandanus leaves through the slits, creating a frame-like design.

2. Add 2 birds-of-paradise at different heights, to the left side of the foam. Use the smaller bird on the top, if they are different in size. Insert 4 Mickey Mouse anthuriums, creating a flower pattern.

3. Create 3 rolled, stapled ti leaves and insert them into the foam in a layered pattern. Add in a picked group of 4 layered ti leaves. See page 13 for foliage techniques.

4. Cut and insert 4 "V"-shaped ti leaves and place them around the anthurium flower cluster. Add in 4 red, wired, ti leaf bows creating a layered pattern. See page 13 for foliage techniques.

5. Add 2 orchid stems cascading to the front right of the design, keeping the front orchid stem shorter then the back stem.

6. Cover the back foam with 2 ti leaf bows and ming fern. Add a touch of ming fern to the front of the design to cover the foam mechanics. See page 13 for foliage techniques.

7. Evaluate the design; adjust, if needed, to complete the "layering" design.

1
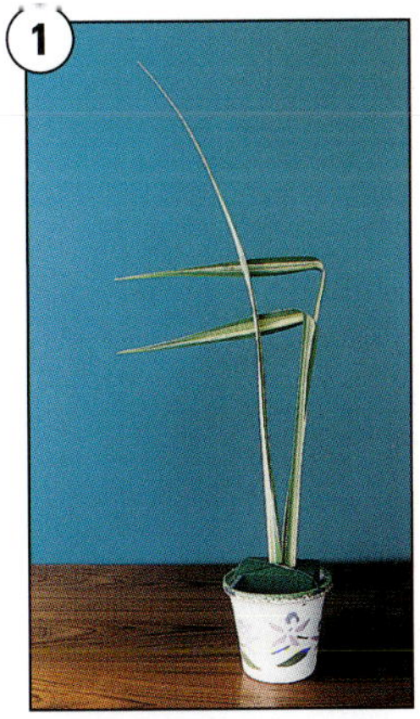

2

3

4

5

6

7

Landscape

MATERIALS:

1 Container – oval with secured, wet foam
1 Cymbidium orchid stem
7 Croton leaves
3 Spider vines
2 Philodendron leaves
3 Ivy stems
1 Bamboo palm flower cluster
2 Ti leaf stalks
4 Brassaia leaves – dwarf
3 Norfolk pine fronds – small tips
2 Sago "palm" husks
1 Sphagnum moss clump

CONTAINER PREPARATION:

Cut a piece of well-soaked floral foam to fit inside the container, leaving about 1/2" for water on both sides of the foam. The foam should extend about 1 1/2" above the top of the container. Secure with waterproof floral tape or hot glue. Add floral preservative and cool water.

The landscape design creates a garden. Floral materials should stimulate the manner in which it grows. Adding pods, branches, rocks, and moss can be used to create such formations as winding paths and boundaries.

1. Insert the cymbidium orchid stem to the back right of the foam. The stem height should be about 2 times the width of the container. Add 7 croton leaves, in a plant formation, in front of the orchid stem.

2. Add 2 spider vines to the front left of the design and 1 inside the design. Insert 2 philodendron leaves to the right side of the design.

3. Insert 3 ivy vines to the left side of the design. Then, add a spike of palm flowers to the center of the design.

4. Add 2 dracena stalks to the left back of the design. Add 4 brassaia leaves and 3 norfolk pine frond tips to the front right of the design, to cover the foam and create interest. Add dried sago "palm" husk in the center of the design for a natural pod-like effect.

5. Add wet, squeezed moss to create a grass-like effect, and to cover the foam mechanics.

6. Evaluate the design; adjust, if needed, to complete the "landscape" design.

1

2

3

4

5

6

New Convention

MATERIALS:

1 Container – medium-deep with secured, wet foam
2 Sexy pink heliconias
4 Pink gingers
3 Mink proteas
5 Safari sunset leucadendrons
2 Laua'e leaves
3 Raphis palm leaves
2 Song of India leaves
1 Sphagnum moss
1 dried branch (any kind)

CONTAINER PREPARATION:

Cut a piece of well-soaked floral foam to fit inside the container, leaving about 1/2" for water on both sides of the foam. The foam should extend about 1 1/2" above the top of the container. Secure with waterproof floral tape. Add floral preservative and cool water. (Optional: add chicken wire before taping down the foam for added support.)

The New Convention technique incorporates strong vertical lines, grouping and clustering of flowers, and adds a bold horizontal plane to the design. It is a combination of a vertical parallel design with a horizontal plane added.

1. Insert 2 sexy pink heliconias, one in front of the other, to the back center of the foam. Create a line with the smallest heliconia on the top.

2. Add 3 gingers to the right, vertical to the heliconia, in a stair-step pattern. Insert 2 mink proteas to the left, vertical to the heliconias.

3. Insert 3 pincushions in a group toward the front of the design, creating the focal point. Add a dried branch to create balance. Insert 2 laua'e leaves to the left and 3 trimmed raphis leaves, 2 next to the pincushions and the other in the back of the design. Add 2 Song of India leaves to the right, creating the horizontal effect of the design.

4. Cover the foam mechanics in the front and back with wet, squeezed moss.

5. Evaluate the design; adjust, if needed, to complete the "new convention" design.

1

2

3

4

5

Colonial

MATERIALS:
- **1 Container – round with secured, wet foam**
- **3 Safari sunset leucadendrons**
- **2 Pincushion proteas**
- **6 Cymbidium orchid blossoms**
- **6 Lady Jane anthuriums**
- **10 Dendrobium orchid sprays**
- **10 Oncidium orchid sprays**
- **2 Ming fern stems**
- **6 4" water tubes**

CONTAINER PREPARATION:

Cut a piece of well-soaked floral foam to fit inside the container, leaving about 1/2" for water on both sides of the foam. The foam should extend about 1 1/2" above the top of the container. Secure with waterproof floral tape. Add floral preservative and cool water.

The Colonial design is compact and round in shape. Flowers are frequently overlapped in the design. At least seven different varieties of flowers must be included in a colonial design, giving the design a new look each time it is viewed. Very little foliage is used in this design.

1. Cut the top safari sunset to about 2 times the width of the vase. Insert it in the center of the foam. Cut the remaining 2 safari sunsets 1" shorter. Add to each side of the centered safari sunsets, creating a dome shape.

2. Add 2 pincushions opposite the 2 shorter safari sunsets. Add 6 cymbidiums in water tubes, 4 on the lower corners of the design and 2 next to the top safari sunset, creating a dome shape.

3. Insert 4 Lady Janes into the bottom of the design, evenly spaced apart. Add 2 Lady Janes next to the top safari sunset, opposite the 2 top cymbidiums, creating the dome shape.

4. Add a touch of ming fern tips to cover the foam mechanics. Insert the dendrobium tips evenly into the design, keeping within the dome shape.

5. Add in the oncidium stem tips, keeping within the dome shape.

6. Evaluate the design; adjust, if needed, to complete the "colonial" design.

1

2

3

4

5

6

Pavè

MATERIALS:

- 1 **Container – low with secured, wet foam**
- 5 **Pincushion proteas**
- 8 **Cymbidium orchids – minis**
- 6 **Safari sunset leucadendrons**
- 7 **Costus gingers**
- 7 **Pink gingers**
- 8 **Red gingers**
- 8 **Epidendrum orchid stems – yellow**
- 2 **Epidendrum orchid stems – purple**
- 5 **Safari sunset leucadendrons – mini**

CONTAINER PREPARATION:

Cut a piece of well-soaked floral foam to fit inside the container, leaving about 1/2" for water on both sides of the foam. The foam should extend about 1 1/2" above the top of the container. Secure with waterproof floral tape. Add floral preservative and cool water.

Pavè is a term borrowed from jewelry-making. The pavè technique "fits" floral materials in place. It is a technique that covers the base of the surface. The surface can be flat, round, or curved as long as the placement height is about the same in depth. It features graceful rows of different types and colors of flowers or materials in different tones and textures.

1. Cut 5 pincushion proteas, leaving 1" stems. Insert them into the foam toward the center, in a diagonal line. Add 8 mini cymbidium orchids next to the pincushions in a diagonal line. Do not leave any space between.

2. Add 6 safari sunsets next to the cymbidium orchids in a diagonal line. Insert 7 costus gingers next to the safari sunsets. Add 7 pink ginger tips next to the pincushions, creating a curved shape.

3. Insert 8 yellow epidendrums next to the costus gingers and 2 purple epidendrums next, in a diagonal line covering the foam mechanics on the side of the vase. Add 8 red gingers and 5 mini safari sunsets in a diagonal line next to the pink gingers. The last row of flowers should cover the foam mechanics on the side of the vase, creating a curved shape.

4. Evaluate the design; adjust, if needed, to complete the "pavè" design.

1

2

3

4

Biedermier

MATERIALS:

- **1 Container – round with secured, wet foam**
- **12 Mickey Mouse Obake anthuriums**
- **6 Safari sunset leucadendrons**
- **17 Cymbidium orchids – mini**
- **3 Leather ferns**
- **1 Pincushion protea**

CONTAINER PREPARATION:

Cut a piece of well-soaked floral foam to fit inside the container, leaving about 1/2" space for water on both sides of the foam. The foam should extend about 1 1/2" above the top of the container. Secure with waterproof floral tape. Add floral preservative and cool water.

The Biedermier design uses the technique of "pavèing and pillowing," but places the flowers in a circular design. The designs have a compact feel and are always round.

1. Insert leather fern tips around and resting on the rim of the container in a circular pattern.

2. Add 8 anthuriums around the vase edge, creating a circle, keeping the fern tips visible.

3. Insert about 17 cymbidiums, in a circular pattern, within the anthurium circle. Keep them close together to cover the foam mechanics.

4. Add in 6 safari sunsets, in a circular pattern, within the cymbidium circle. Keep them close together to cover the foam mechanics.

5. Insert the pincushion in the top center to conclude the cake-like effect.

6. Evaluate the design; adjust, if needed, to complete the "Biedermier" design.

1

2

3

4

5

6

Flemish

MATERIALS:

- **1 Container – tall with preservative and cool water**
- **10 Dendrobium orchid sprays**
- **12 Mickey Mouse Obake anthuriums**
- **5 Parrot heliconias**
- **7 Safari sunset leucadendrons**
- **6 Leather ferns**
- **4 Plumosa ferns**
- **7 Spreingeri ferns**

Flemish bouquets recreate the floral paintings by the Old Dutch masters. Flowers at various stages of development, from baby buds to blossoms past their prime, and a diversity of flower forms and sizes, are combined in Flemish bouquets. The bouquet is an all-around design. A flowing, draping look is associated with this design. This design uses the "lacing" technique by lacing the stems crossing each other to support the flowers without the use of floral foam, chicken wire, or a needle pin holder.

1. Clean and insert 4 plumosa ferns in a cris-cross pattern to create a grid to lock the stems in place. Create an all-around bouquet. Be sure the stems are especially clean when using a clear glass vase. Strip off all foliage and flowers that may fall below the water line. Add in 6 leather ferns, cris-crossing the stems, creating an added gridlock. Add 7 springerai ferns, cris-crossing the stems, creating a cascading effect to the right side.

2. Add 5 dendrobiums in a criss-cross pattern along the rim of the vase at an even distance apart. Insert the remaining 5 dendrobiums toward the top, creating the height of the bouquet in a triangular form. Remember to strip off the lower flowers.

3. Measure and cut the top parrot 2 1/2 to 3 times the height of the vase. Cut the remaining 4 parrots 2 1/2" shorter than the top parrot. Insert the longest parrot into the top center of the bouquet. Add in the 4 parrots around the top parrot.

4. Insert 12 anthuriums evenly, around the bouquet, at equal distances apart. Start at the rim of the vase, upward to the top, center. Space the anthuriums within the triangular form.

5. Add in 7 safari sunsets, evenly spaced apart, between the anthuriums. Keep within the triangular form. Remember to strip off the lower leaves.

6. Evaluate the design; adjust, if needed, to complete the "Flemish" design.

1

2

3

4

5

6

Hand-Tied

MATERIALS:

1 Container
4 Pink gingers
7 Dendrobium orchid sprays
4 Marion Seefurth anthuriums
2 Curly willow branches
6 Laua'e leaves
14 Bear grass stems
1 Monstera leaf
2 Bamboo orchid stems
4 Pampas grass stems
2 Raffia or twine (yards)

TIPS:

- Start with your flowers and foliage laid out on the table.
- Remove the flowers and foliage from the lower half of all stems.
- If you are right handed, hold the flowers in your left hand.
- Hold the flowers and foliage halfway up the stems, between your thumb and index finger.
- Start with flowers that have a vertical line.
- Place similar colors across from one another to create a burst of color and balance.
- Balance is critical.
- Keep your hands relaxed.
- Do not squeeze your flowers too tightly.
- If holding the bouquet until it is finished presents a problem, it can be tied up at intervals during the arranging process.
- Do not use wire to tie.

The hand-tied technique is designed in the hand, in a crossing pattern, and then securely tied. A hand-tied bouquet will stand alone on its trimmed stems before being dropped into a vase filled with water. Hand-tied bouquets should include a wide variety of flowers, along with several different types of foliage.

1. Start with 1 ginger in your left hand and one laua'e in your right hand. Place the laua'e on the bottom at a 45-degree angle.

2. Add 1 laua'e and 1 ginger at a 45-degree angle, building your bouquet in a clockwise direction.

3. Add 1 low ginger, 1 laua'e, and 1 dendrobium, continuing in the clockwise direction. Turn the bouquet minutely with your thumb as the flowers are added. Keep the stems crossing at a 45-degree angle.

4. Add 1 ginger, 1 dendrobium, and 1 curly willow branch at a 45-degree angle, in a clockwise direction. Turn bouquet minutely with your thumb as the flowers are added.

5. Add 1 laua'e, 1 curly willow, 1 dendrobium, and 1 bamboo orchid stem at a 45-degree angle, continuing in the clockwise direction. The flowers or foliage that curve outward determine the width of the bouquet.

6. Add 1 anthurium, half of the bear grass, 1 bamboo orchid stem, 1 dendrobium, and 1 monstera leaf, continuing in the clockwise direction.

7. Add 1 anthurium, 1 laua'e, 1 dendrobium, the remaining bear grass, and 1 pampas grass. Add 1 anthurium, 1 dendrobium, 1 pampas grass, the remaining bear grass, and 1 laua'e. Then, add 1 anthurium, 1 dendrobium, 1 pampas grass, and 1 laua'e. Add 1 anthrium, 1 dendrobium, and 1 pampas grass. Lastly, add 1 dendrobium, 1 pampas grass, and 1 monstera leaf keeping them all in the clockwise direction in a 45-degree angle. Add in all the flowers and foliage until a desired look and size is reached.

8. Bind the stems together with raffia at the point where your hand is holding the bouquet. Wrap the raffia around the stems as many times as necessary to hold the bouquet firmly in place before tying.

9. Secure the binding by pulling the raffia up through the stems and back around the bind, tugging gently. Continue this step 2 to 3 times. Tie a knot to secure the raffia.

10. Trim stems evenly. The bouquet should be freestanding, ready to be placed in a vase. A well-balanced bouquet should be able to stand by itself on its stems.

11. Evaluate the design; adjust, if needed, to complete the "hand-tied" design.

1

2

3

4

5

6

7

8

9

10

11

ORIENTAL *Style* ARRANGEMENTS

Ikebana means "The Way of the Flower." It is a traditional Japanese floral art that recreates the impression of a plant's beauty, choosing the most beautiful aspects of the materials used. Ikebana arranges plants that are cut and removed from nature and filled with new beauty when placed in a new environment. It is usually a display of strong lines and a sparing use of flowers creating a clean, simple look, unlike the American and European floral arrangements. An asymmetrical form and the use of empty space is an important feature of the design.

Ikebana is based upon the principles of dominance, proportion, balance, and rhythm to symbolize the continued concepts of Buddhism from China nearly 1,500 years ago. The form of this design involves three elements consisting of lines representing heaven, man, and earth—each of which is fundamental to the design. The primary line is the tallest. It is called Shin, which represents heaven. A secondary line, which represents man, is called Soe. A tertiary line is called Tai or Hikae, which represents earth. Filler flowers, or helpers, called Aishidai or Jushi, are used to tie together the three main elements in many ikebana arrangements.

Ikenobo is said to be the origin of ikebana. Subsequently, other schools have branched off from Ikenobo. The three leading schools of ikebana in Japan are the Ikenobo School, the Sogetsu School, and the Ohara School. The study of ikebana involves taking the path that your teacher has walked by learning over many years the essence of nature and the human spirit. For this book, a simple and short approach is shown.

This section shows seven different types of arrangements such as Shoka, Rikka, Heika, Moribana, Juuka, and two different styles of Nageire. Every school follows different principles. Following are a few basic guidelines in Japanese flower arranging.

Basic Guidelines *in Japanese Flower Arranging*

1. Use one dominant material plus intermediate and subordinate materials.

2. Place materials to appear as if it rises from a central point.

3. Show the curves of a stem to their best advantage.

4. Use different heights to give individuality to each stem.

5. Make sure the arrangement and all of its parts are well balanced or asymmetrically balanced.

6. Check to see if everything is in harmony and the whole arrangement is in unity.

7. Allow the tips to lift and turn slightly forward.

8. Have a definite focal point.

9. Use foliage in proper relationship to blossoms.

10. Use more foliage than blossoms.

11. The total number of flowers are usually odd.

12. Practice restraint.

13. Avoid duplication, crossing, and a parallel pattern of lines.

14. Allow no flower or foliage to touch the water or the container.

15. Cut the stems at a 45-degree right angle if the stem is to lean to the left in the frog.

16. Cut the stems at a 45-degree left angle if it is to lean to the right in the frog.

17. Cut the stems straight across for an upright placement into the frog and for a thin stem.

18. Cut the stems under water in a bowl of tap water to prevent stress on the plant materials, especially flowers.

Shoka

MATERIALS:

- **1 Container – pedestal with pebbles (if needed)**
- **1 Needle pin holder (frog or Kenzan) – large**
- **2 Bird-of-Paradise flowers**
- **7 Bird-of-Paradise leaves**

Shoka or Seika means "living flowers." It is a simplified form of Rikka. Shoka rises gracefully from the water's edge. It is an impression of a plant's simplicity and beauty. Shoka illustrates the process of growth: a seed is planted, it sprouts, it grows toward the sun, it is affected by the weather, and it eventually wilts and dies. The simple nature of its three main branches is Shin for heaven, Soe for man, and Tai for earth, which forms a unity that expresses life's perpetual change and renewal. The final part, Ashidai (helper), is often added as a finishing touch.

The footing of Shoka is a back-to-front linear pattern arranged in the frog (Kenzan) with a few stems placed next to the stems in the main line. If many stems are used, the footing forms a triangle. Shoka Shofutai is the traditional form, with rigid rules. Shoka Shimputai is a new style with a bright, modern feeling. Demonstrated is a Shoka Shimputai Ishuike (one type of material).

SHOKA *Continued from previous page.*

1. Measure and cut the best, straightest bird-of-paradise to two times the height and the width of the container. Insert it into the middle of the frog. Cut 3 bird-of-paradise leaves: Cut the first leaf about 4" shorter than the stem of Shin; cut the second leaf about 4" shorter than the first leaf; cut the third leaf about 6" shorter than the second leaf. Insert the long leaf next to the Shin stem. Insert the medium leaf in back of the Shin stem. Finally, insert the shortest leaf in front of the Shin stem. All leaves should be facing the flower (Shin). This completes the Shin group.

2. Measure and cut a bird-of-paradise leaf three-fourths the length of the bird-of-paradise (Shin). Insert it in back of the Shin stem facing the flower (Shin). This completes the Soe group.

3. Measure and cut the last bird-of-paradise flower (Tai) to about one-half the length of the first flower (Shin). Insert the stem into the frog in front of the Shin group. Measure and cut 3 additional bird leaves. Cut and insert the longest leaf about 1" shorter than the shorter flower (Tai), and insert it to the right of the shorter flower. Cut 2 more bird leaves, the first about three-fourths the length of the long bird leaf, and the second about half the length of the long bird leaf. Insert the shorter of the 2 leaves in front of the short flower, flowing forward to cover the stem mechanics. Insert the longer leaf to the left of the flower to cover the side mechanics. This completes the Tai group.

4. Evaluate the design; adjust, if needed, to complete the "Shoka Shimputai Ishuiki" (one type of material) arrangement.

1

2

3

4

Rikka

Rikka means "standing flowers." It describes arrangements which were reproductions of those made by Buddhist priests for their altars. It is usually very formal. These arrangements portray a scene in the natural landscape. A sense of the structure is one of the most important factors to consider. The object of Rikka is to create a symbolic representation of the cosmos using 5, 7, or 9 different plant materials. The footing of Rikka is created in a cluster, forming a circle. The circle signifies peace. There is the classical Rikka (Shofutai), and the modern Rikka (Shimputai). The modern-style Rikka evolved from homes having a contemporary environment. Modern Rikka permits freedom of choice with materials, omission of principle parts, and the use of modern containers.

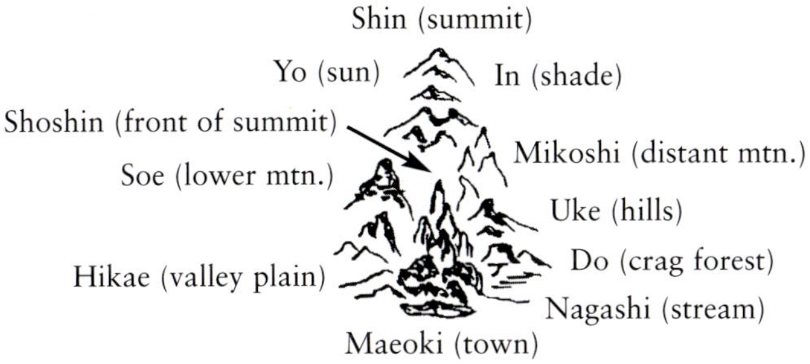

Rikka consists or 9 basic parts arranged into a single vase to express the beauty of a natural landscape. The basic parts are the following:

Shin (summit): tallest and strongest,

Shoshin (front of summit): straight material, two-thirds to one-half of Shin,

Soe (lower mountain): front side faces Shin, about 2/3 of Shin,

Uke (hills): shorter than Soe,

Mikoshi (distant mountain): shorter than Shin and longer than Soe; lightweight material,

Hikae (valley plain): back of Soe: extend to the rear,

Nagashi (stream): its length is the same as Soe, footing across of Hikae to the right,

Do (crag forest): the body or thickest portion of Rikka: shorter than Shoshin,

Maeoki (town): smaller mass placed in front of Do; its backside shows and is bent forward at the water level.

Plus:

Ushiro Gakoi (back wall): a plant material placed in the back of Shin to create depth. It is the finishing touch.

Modern Rikka (Shimputai)

MATERIALS:

- **1 Container – usually a compote or pedestal**
- **1 Needle pin holder (frog or Kenzan) – large**
- **5 Wires – #22**

7 PLANT MATERIALS:

- **20 Pandanus leaves**
- **2 Areca palm fronds**
- **4 Parrot heliconias**
- **3 Anthuriums**
- **2 Oncidium orchids**
- **3 Ferns**
- **4 Gardenia branches**

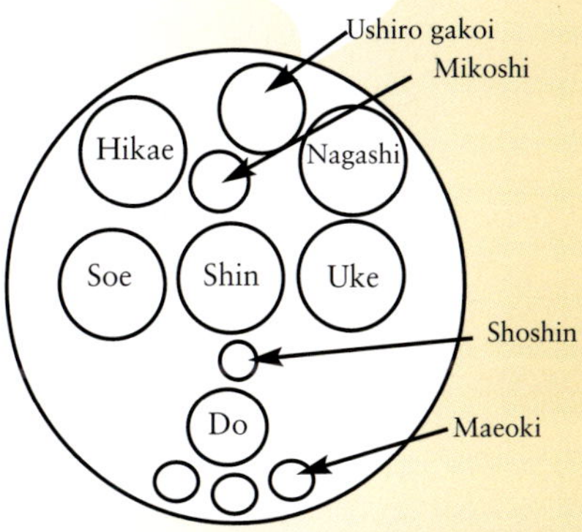

1. Cut 3 pandanus leaves at different lengths with the tallest, straightest leaf to two times the height and width of the container. Place the 3 pandanus leaves at different lengths in your hand, with the shortest leaf in the front. Trim to even the bottom of the stems creating a cluster. Insert all three leaves into the frog slightly behind the middle point. Trim the straighter palm frond to the same length as the tallest pandanus leaf. Insert the stem in the back of the pandanus cluster of leaves. This completes the Shin group of the arrangement.

2. Cut the last palm frond to three-forths the length of the tallest pandanus leaf in the Shin group. Insert the stem to the left of the Shin group with the palm frond leaning to the left, balanced with the first palm frond in the Shin group. Add 2 pandanus leaves as helpers (Aishidai) to the palm frond to the left (Soe). Trim the pandanus leaves to shorten them. Place the taller, trimmed leaf in back of the palm frond stem and a shorter leaf in front of the palm frond stem. This completes the Soe group.

3. Cut the best heliconia (Shoshin) to half the length of the tallest pandanus leaf in the Shin group. Insert it into the frog in front of the Shin group. Add a second heliconia (Aishidai) as front helper to the first heliconia (Shoshin). Add a third heliconia (Aishidai) as a back helper in back of the first heliconia (Shoshin).

4. Insert 2 clusters of shortened pandanus leaves (Do) at the center and extending to the left and right of the arrangement.

5. Cut an anthurium (Uke) shorter than the palm frond to the left (Soe). Support the anthurium with 1 longer, trimmed pandanus leaf underneath and 1 shorter, trimmed pandanus leaf behind it. Wire the stems together for support and stability. Bend to a 45-degree angle. Insert into the frog on the right side of the Shin group.

MODERN RIKKA (SHIMPUTAI) *Continued from previous page.*

6. Trim and cluster with wire 2 pandanus leaves and 1 heliconia to a length shorter than the anthurium (Uke) group. Bend to a 45-degree angle to the left. Place it into the frog in front of the Soe group to balance the anthurium group (Uke).

7. Trim and cluster with wire 2 pandanus leaves to a length shorter than the anthurium cluster (Uke). Bend to a 45-degree angle to the right. Insert it into the frog in back of the anthurium cluster (Uke).

8. Add 2 anthuriums to the front of the arrangement; one longer extending up and forward, and the other shorter and lower. This completes the Do area, creating brightness and depth.

9. Add 2 stems of orchids to create additional brightness and depth. Place a shorter orchid into the frog, to the left of the arrangement, behind the palm frond (Soe). Place a longer orchid (Mikoshi) into the frog behind the Shin group, to the right of the arrangement. See step #10 for stem reinforcement.

10. Bind to reinforce the orchid stem with an anthurium stem and wire if added support is needed for insertion into the frog.

11. Add 1 trimmed gardenia branch (Nagashi) to the right of the arrangement at a 90-degree angle with the tip of the leaves facing upward. Place the stem into the frog next to the tall orchid stem (Mikoshi) and behind the anthurium to the right (Uke). Add 3 additional trimmed gardenia branches (Maeoki) in the front, and to the right and left side of the forward anthurium (Do). Keep the stem area and water line clear of leaves.

12. Add 3 ferns (Hikae) to the left side of the arrangement, at different lengths. Place the longest fern in the middle, the shortest fern in the back, and the last fern in the front. See step #13 for stem reinforcement. Although not visible in the photo, a pandanus leaf slightly shorter than the tall heliconia (Shoshin) has been added in the back, as a back wall (Ushiro Gakoi) for depth. It is positioned leaning slightly toward the back.

13. Reinforce the fern stem by capping it with an anthurium stem if added support is needed for insertion into the frog.

14. Evaluate the design; adjust, if needed, to complete the "Rikka Shimputai" (modern) arrangement.

1

2

3

4

5

6

7

8

9

10

11

12

13

14

Heika

Heika means "vase flowers." It is a basic style of ikebana that uses the same principles as Rikka and Shoka, eliminating some of their points. It is arranged in a tall, deep container with a narrow opening. Crosspieces are used, if support is needed, to keep the stems in place. A Heika arrangement consists of three main elements – the subject or primary, the secondary, and the object stem along with their helpers, called intermediaries. Their lengths, positions, and angles are different, depending on the Heika style selected. There are three variations of basic Heika forms: the slanting style, the upright style, and the cascading style. Demonstrated is a cascading-style Heika arrangement.

MATERIALS:

1 Container – tall with a narrow opening, and branch fixture (optional)
1 Phalaenopsis orchid spray
3 Song of India leaves
2 Plumosa ferns

1. Place the phalaenopsis orchid spray (subject) cascading from the front left of the rim of the vase.

2. Insert a tall Song of India leaf (secondary) to the back right of the design.

3. Place a short leaf (intermediary) backward, in front of the tall leaf. Trim the leaf from the base to shorten if needed.

4. Trim the last Song of India leaf (object) to about half the length of the tall leaf. Insert it into the right side of the design.

5. Add 2 plumosa fern tips (intermediaries) to the front right and back left of the arrangement.

6. Evaluate the design; adjust, if needed, to complete the "cascading-style Heika" arrangement.

1

2

3

4

5

6

Moribana

Moribana means "a mounding or piling up of materials." It was developed due to the western influence on Japan. It was created for homes furnished in the western-style manner. Moribana is arranged in a shallow vase called a "Suiban," or water basin, with the help of a needle pin holder called a "Kenzan," or frog. Moribana consists of three main groups: Shin (heaven), Soe (man), and Hikae (earth), along with their Jushi (helpers). The stems are placed into the frog in a harmonized triangle shape. Moribana consists of 4 elements: line, space, color, and mass. There are 8 frog positions. The frog is never placed in the middle of the container in Moribana. There are 8 variations of basic Moribana arrangements. Demonstrated is a basic slanting-style Moribana arrangement.

MATERIALS:

- **1 Container – low, shallow (Suiban) with pebbles (if needed)**
- **1 Needle pin holder (frog or Kenzan) – large**
- **1 Pebble bag**
- **1 Strawberry guava main branch – curving to the left at 45 degrees**
- **1 Strawberry guava smaller branch – curving to the left at 45 degrees**
- **3 Strawberry guava branches**
- **5 Red gingers**

1. Select the main guava branch (Shin) to lean at 45 degrees toward your left shoulder. Cut it to the length of the (diameter of vase + depth of vase) x 2. Trim off conflicting leaves and branches. Hold the stem in the exact angle it is to be positioned and re-cut it at a 45-degree angle on the right. This angle secures the branch when leaning to the left.

2. Place the frog in the back right of the container (position #3).

3. Position the branch toward the front left of the frog at 45 degrees toward your left shoulder. Trim off conflicting leaves and branches.

4. Select a straight guava branch for Soe. It should be three-fourths the length of the Shin branch. Trim off conflicting leaves and branches. Hold the stem in the exact angle it is to be positioned and re-cut it at a 45-degree angle on the right. This angle secures the branch when leaning to the left. Insert Soe into the back center of the frog, creating the back point of an imaginary triangle. Position it to lean at 15 degrees toward the left. Trim off conflicting leaves and branches.

5. Select the best ginger for Hikae. Cut it to half the length of the Soe guava stem at a 45-degree angle to the left. This angle secures the branch when leaning to the right. Slowly soften and bend the stem if needed to position it at 75 degrees toward your right shoulder. Insert it into the frog on the front right, creating the right point on an imaginary triangle.

6. Insert the 4 remaining gingers into the frog as helpers (Jushi) to the main ginger (Hikae). The helpers should be of different lengths, kept within the 75-degree position, inserted next to the ginger (Hikae), while leaning toward the right shoulder. The flowers should point up to the sun and not touch the water or the rim of the vase. This completes the Hikae section.

7. Trim off the conflicting leaves and branches of the smaller 45-degree curved guava branch. Cut it to a length shorter than the main branch (Shin) to add as a helper (Jushi) in the back of the main branch (Shin). (Make sure it is shorter than the Soe branch.) Hold the stem in the exact angle it is to be positioned and re-cut the stem at a 45-degree angle to the right. Insert it in back of the Shin stem at a 45-degree angle leaning to the left. It should complement the main stem.

MORIBANA *Continued from previous page.*

8. Select a front helper stem (Jushi) to be placed in front of the Shin stem. It should be shorter than the back helper (Jushi) to Shin, but long enough to flow forward and cover the stem mechanics of the design. This front helper (Jushi) gives the illusion that the guava branches have become a growing plant.

9. Select and trim off the conflicting leaves and branches for a back helper (Jushi) to Soe. It should be about half the length of Soe. Insert it in back of Soe, slightly to the left creating the back depth.

10. Cover the frog mechanics with a touch of small guava branch tips, keeping the leaves clear from the water level. Pebbles may be used, as a substitute, to cover the frog mechanics.

11. Evaluate the design; adjust, if needed, to complete the basic "slanting Moribana" arrangement.

1

2

3

4

5

6

7

8

9

10

11

Nageire

Nageire means thrown or tossed in. The arrangements have a natural and informal feeling of simplicity. Nageire is usually arranged in a tall, narrow-mouth container without the use of a frog (Kenzan). The mouth of the container should be clearly defined in a Nageire-style arrangement. The stems are braced against the inside of the vase to hold the flowers in place. Supportive fixtures and other techniques are used to hold the stems in a desired position. Fixtures such as the crossbar (Jumonji-Dome), the vertical fixture (Soegi-Dome), and the direct fixture (Jika-Dome) are used. A hibiscus branch, about 3/4" thick, is recommended because its flexible, fibrous, strong stem is needed while creating the crossbar (Jumonji-Dome) fixture. A Nageire arrangement consists of three main parts: Shin, Soe, and Hikae. Shin (heaven) is the dominant branch; Soe (man) is the subdominant branch; and Hikae (earth) is a subordinate branch. Nageire consists of the 4 elements: line, space, color, and mass. In Nageire, a flexible attitude toward your materials enables you to adjust if the indicated angle of the main stem is off in position. There are six variations of basic Nageire arrangements. Demonstrated are the basic upright- and slanting-style Nageire arrangements.

Basic Upright Style Nageire

MATERIALS:

1 **Container – tall, cylinder-shaped**
1 **Hibiscus branch (or similar) – 3/4" thick**
4 **Mock orange stems**
3 **Parrot heliconias**

Upright-style Nageire arrangements consist of stems forming lines that extend upward from the container. The main branch, Shin, should be strong and stable. Bend the branch as much as flexibility allows for a softer, flowing design. It is similar to the basic upright Moribana-style arrangement, but it uses a tall cylinder vase with a crossbar, branch fixture (Jumonji-Dome) to secure the stems instead of a frog (Kenzan). A thick, sturdy container is best when applying the crossbar (Jumonji-Dome) fixture method.

1. Cut a piece of the hibiscus branch just a bit longer than the inside diameter of the vase. Cut one end straight and the other end at a 45-degree angle.

2. Place the stem inside the vase with the flat side of the stem about 2" below the rim of the vase. Position the 45-degree slant facing upward, about 3" below the rim of the vase. Slowly, pull up the branch creating the horizontal fixture.

3. Cut another piece of the hibiscus branch just a bit longer than the inside diameter of the vase. Cut one end straight and the other end at a 45-degree angle. Position it across the first branch with the 45-degree slant cut facing downward. Push down to create a cross. If it is unstable, repeat steps #1 through 3.

4. Select a strong and stable mock orange branch (Shin) that is straight with a slight curve to the left. If the branch is not naturally curved, gently bend the branch as much as flexibility allows creating a slight curve. Trim off conflicting leaves and branches. Measure and cut the stem to (the diameter of the vase + the depth of the vase) x 2. Hold the stem in the exact angle it is to be positioned and re-cut it at a 45-degree angle on the right. This angle secures the branch against the vase, when leaning to the left.

5. Place the mock orange stem (Shin) into the back right quadrant, leaning toward your left shoulder.

6. Lean the stem (Shin) toward your left shoulder at a 10- to 15-degree position. Trim off conflicting leaves and branches, keeping this main stem (Shin) strong.

7. Select another mock orange branch (Soe). Trim off conflicting leaves and branches. Measure and cut it to three-fourths of the length of the tall mock orange branch (Shin). It should be thinner and lighter than the tall mock orange branch (Shin). Hold the stem in the exact angle it is to be positioned and re-cut it at a 45-degree angle on the right. This angle cut secures the branch against the vase when leaning to the left.

8. Place the mock orange stem (Soe) into the back right quadrant, along with the Shin stem, leaning toward your left shoulder. It may be placed in front or the back of the tall mock orange stem (Shin). Trim off the conflicting leaves and branches.

BASIC UPRIGHT-STYLE NAGEIRE *Continued from previous page.*

9. Select the best curved-stem heliconia (Hikae). Gently bend the stem as much as flexibility allows. Keep your thumbs close together and hands by your navel while bending the branch.

10. Measure and cut the heliconia stem (Hikae) half the length of the shorter mock orange branch (Soe). Place it into the front left quadrant, leaning toward your right shoulder at 75 degrees from the top.

11. Add 2 heliconias as helpers (Jushi) to the main heliconia (Hikae). Insert them into the same front left quadrant as the main heliconia (Hikae), leaning toward your right shoulder. Keep them within the 75-degree position at different lengths. Place one in the front and the other in the back, leaning forward.

12. Select a thin branch that curves to the left. Trim off the conflicting leaves and branches. Cut it to three-fourths the length of the left branch (Soe). Place it in front of the Soe branch, as a front helper (Jushi), to complement it. Select a straight, thin mock orange branch (Jushi) as a back helper to the main branch (Shin). Trim off the conflicting leaves and branches. Measure and cut the branch half the length of the branch on the left (Soe). Hold the stem in the exact angle it is to be positioned and re-cut it at a 45-degree angle toward the front. This angle cut secures the branch against the vase when it leans toward the back. Insert the mock orange branch as a helper (Jushi) into the back quadrant, behind the main branch (Shin), leaning to the left. This creates back depth.

13. Select a thin branch (Jushi) for a front helper to Shin. Cut it shorter than the back helper (Jushi) to the Shin branch. Insert it into the back right quadrant in front of the Shin branch as a front helper (Jushi). Trim off the conflicting leaves and branches.

14. Select a thin branch as a front helper (Jushi) to the Soe branch. It should be shorter than the main Soe branch, but longer than the front helper (Jushi) to the Shin branch. Place it into the back right quadrant in front of the Soe branch, leaning towarrd your left shoulder. Trim off the conflicting leaves and branches.

15. Evaluate the design; adjust, if needed, to complete the basic "upright Nageire" arrangement.

1

2

3

4

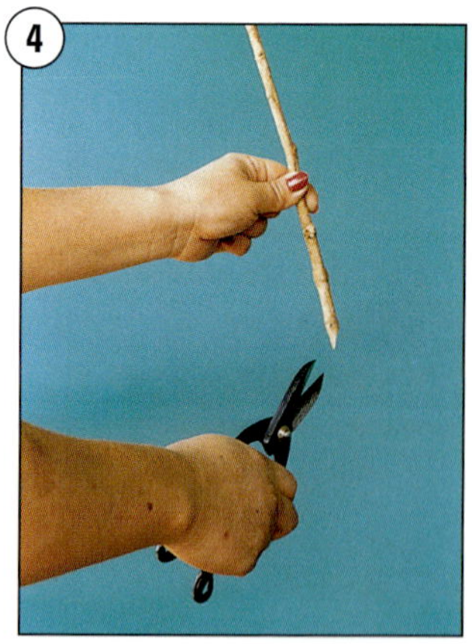

5

6

7

8

9

10

11

12

13

14

15

Basic Slanting Style Nageire

MATERIALS:

1 **Container – tall, cylinder-shaped**
1 **Hibiscus branch (or similar) – 3/4" thick**
1 **Teakwood branch – tall, curving upward to the left**
2 **Teakwood branches – straight**
5 **Parrot heliconias**

Slanting-style Nageire arrangements consist of stems forming lines that extend on a slant from the container. Choose branches that look beautiful when slanted and have a light and vigorous linear expression. This style gives a softer impression compared to the upright-style Nageire arrangement.

1. Cut a vertical fixture (Soegi-Dome) from a hibiscus branch to the height of the vase, less 1". Both ends should have a straight cut. If the fixture is too tall, the mechanics will show.

2. Cut the top of the fixture stem in half, equally centered, about 2 1/2".

3. Select a Shin branch that is curved to the left, facing forward and up. Trim off the damaged leaves and conflicting leaves and branches, keeping the main stem line clean and strong.

4. Measure the height of Shin to (the diameter of the vase + the depth of the vase) x 1.5. Hold the branch in the exact position it will be placed, and cut a slit equally in half, about 2" long on the side of the stem.

5. Re-cut the end of the Shin branch to a 45-degree angle with the point of the cut toward the bottom. This will allow the stem to firmly lean against the inside wall of the vase. Interlock the Shin branch onto the vertical fixture. Place it into the vase with the Shin branch leaning toward your left shoulder at 45 degrees from the top.

6. Select and trim off the damaged leaves and conflicting leaves and branches of a straight Soe branch. Keep the main stem line clean and strong.

7. Cut the Soe branch three-fourths the length of the Shin branch. Hold the Soe branch facing forward, and cut a vertical slit, equally in half, about 2 1/2" long.

8. Interlock the Soe branch with the Shin branch.

9. Insert the branches anchored to the fixture into the vase. Position the Shin branch at a 45-degree angle to your left shoulder. The Soe branch should be about three-fourths the length of the Shin branch, 15 degrees to the left. Take apart and make adjustments if needed. Trim off any conflicting leaves and stems that distract from the flow of the line on the branch.

10. Measure and cut the best, curved-stemmed heliconia (Hikae) to three-fourths the length of the Soe. Slowly bend the stem with both thumbs close together to create a 75-degree position when inserted into the vase. Hold it at the exact position to be placed and re-cut the stem at a 45-degree angle to the left. Add the heliconia (Hikae) facing your right shoulder, at a 75-degree position. It may or may not be tucked into the fixture.

11. Add 4 additional shorter heliconias as helpers (Jushi) to the first heliconia (Hikae) at different lengths, leaning toward your right shoulder. This completes the Hikae section of the arrangement.

12. Add a shorter, trimmed branch in front of the Shin branch as a helper (Jushi) to strengthen the Shin branch if it is needed. It must be shorter than the Soe branch and not distract from the main, strong line of the Shin branch.

13. Trim off the conflicting leaves and branches, keeping the main stem line clean and strong.

14. Evaluate the design; adjust, if needed, to complete the "slanting Nageire" arrangement.

1

2
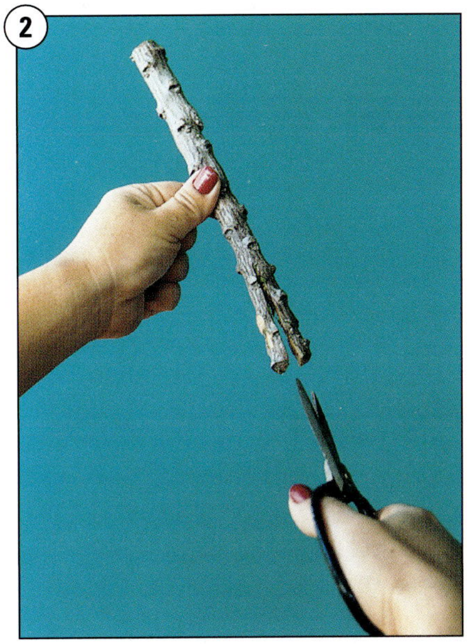

3

4

5

6

7
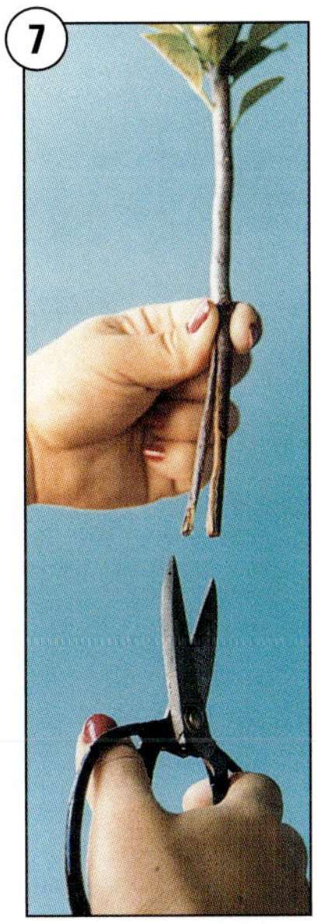

8

9

10

11

BASIC SLANTING-STYLE NAGEIRE *Continued from previous page.*

(12)

(13)

(14)

Juuka or Jiyuka

MATERIALS:
1 **Container – with secured, wet foam, branch or chicken wire fixture (optional)**
1 **Phalaenopsis orchid spray**
2 **Pandanus leaves**
2 **Song of India leaves (or similar)**
1 **Epidendrum orchid stem**
1 **Asparagus fern stem**
2 **Wire – #22 (or similar)**

Juuka means "free style." Juuka is the most recent form of Ikenobo ikebana to emerge from personal expression. It is a very modern approach to the art, suited to contemporary environments and tastes. Juuka is a freestyle-type of arrangement that reshapes its container. Juuka incorporates a mystery flower that is of a bright color that peeps out from the back. It is a forward-facing arrangement without back depth. Juuka is divided into a naturalistic and an abstract style. Both styles use plant materials in a new way, giving respect to the beauty and qualities of each material. Juuka's possibilities are endless, flowing from the arranger's creativity in using the materials. Demonstrated is an abstract style.

1. Cut and insert the phalaenopsis orchid spray (Tai) toward your right shoulder. Place the spray following the curve of the container, not allowing it to touch the table.

2. Cut the straightest pandanus leaf (Shin) the length of about two times the height and the width of the container. Cut at a 45-degree angle from the spine of the leaf outward. Secure with wire to strengthen the inserting tip.

3. Cut the second pandanus leaf (Soe) at a 45-degree angle, from the spine of the leaf outward. Cut the leaf to about three-fourths the length of the first leaf (Shin). Pinch, bend, and secure with wire to create a 45-degree angle and strengthen the inserting tip.

4. Insert the longest pandanus leaf (Shin) into the foam, straight up, following the shape of the neck of the container. Insert the second pandanus leaf (Soe) bending downward to the left, reshaping the container. It should not touch the table. Add 2 Song of India leaves diagonally on each side of the arrangement, keeping the right side shorter in length.

5. Add in the epidendrum orchid in back of the phalaenopsis orchid spray, just peeping through. Cover the mechanics with a touch of ming fern if foam is used.

6. Evaluate the design; adjust, if needed, to complete the "Juuka" arrangement.

1

2

3

4

5

6

BIBLIOGRAPHY

Anderson, Joyce Y. & Fleetham, B. Joan. *Professional Floristry Series: Advanced Techniques.* British Columbia: Lexicon Enterprises Ltd., 2000.

Berry, Fred & Kress, John. *Heliconia: An Identification Guide.* Washington, DC: Smithsonian Institution Press, 1991.

Brenzel, Kathleen Norris. *Western Garden Book.* Menlo Park, CA: Sunset Publishing Corp., 2001

Broschat, Timothy K. & Meerow, Alan W. *Betrock's Reference Guide to Florida Plants.* Hollywood, FL: Betrock Informations Systems Inc., 1994.

California Protea Association. *California Protea: Exotic Protea Flowers for the Retail Florist.* Valley Center, CA: http://www.californiaprotea.com/index.html, 1998.

John Henry Company. *Basic Floral Design.* Lansing, MI: John Henry Company, 1991.

Kepler, Angela K. *Exotic Tropicals of Hawaii.* Kailua, HI: Press Pacifica, Ltd., 1999.

Leonhardt, Ken & Sewake, Kelvin eds. *Growing Dendrodium Orchids in Hawaii: Production & Pest Management.* CTAHR, University of Hawai'i. Honolulu, HI: 1999.

Matthews, Lewis J. *The Protea Book: A Guide to Cultivated Proteaceae.* Portland, OR: Timber Press, 2002.

Neal, Marie C. *In Gardens of Hawaii.* Honolulu: Bishop Museum Press, 1965.

Robelo, Tony. *Sasol Proteas: A Field Guide to the Proteas of Southern Africa.* Vlaeberg, South Africa: Fernwood Press, Association of National Botanical Institutes, 1995.

Shinno, Tat. *Flower Arranging by Tat.* Menlo Park, CA: Lane books, 1965.

Stark, Randall E. Sr. *A Complete Guide to Retailing Fresh Flowers.* Fullerton, CA: Elva May Florist School, 1984.

Takahashi, Mary C. *Ikenobo Ikebana.* Kyoto, Japan: Kawakita Printing, 1974.

Upcountry Harvest Gift Shop. *Flower Facts.* Kula, HI: http://www.upcountryharvest.com/index.html, 2003.

Wagner, Warren L., Herbst, Derral R., & Sohmer, S. H. *Manual of the Flowering Plants of Hawaii, Volume I & II.* Honolulu: University of Hawai'i Press, Bishop Museum Press, 1990.

Whistler, W. Arthur. *Tropical Ornamentals – A Guide.* Portland, OR: Timber Press, 2000.

Wood, Paul. *Proteas in Hawaii.* Waipahu, HI: Island Heritage Publishing, 2003.